MICROSCOPY HANDBOOKS 26

Enzyme Histochemistry.
A Laboratory Manual of Current Methods

Cornelis J. F. Van Noorden
and Wilma M. Frederiks

Laboratory of Cell Biology and Histology
University of Amsterdam
Academic Medical Centre
Meibergdreef 15
1105 AZ Amsterdam
The Netherlands

Oxford University Press · Royal Microscopy Society · 1992

Oxford University Press, Walton Street, Oxford OX2 6DP
Oxford New York Toronto
Delhi Bombay Calcutta Madras Karachi
Kuala Lumpur Singapore Hong Kong Tokyo
Nairobi Dar es Salaam Cape Town
Melbourne Auckland Madrid
and associated companies in
Berlin Ibadan

Royal Microscopical Society
37/38 St Clements
Oxford OX4 1AJ

Oxford is a trade mark of Oxford University Press

Published in the United States
by Oxford University Press Inc., New York

© *Royal Microscopical Society, 1992*

All rights reserved. No part of this publication may be
reproduced, stored in a retrieval system, or transmitted, in any
form or by any means, without the prior permission in writing of Oxford
University Press. Within the UK, exceptions are allowed in respect of any
fair dealing for the purpose of research or private study, or criticism or
review, as permitted under the Copyright, Designs and Patents Act, 1988, or
in the case of reprographic reproduction in accordance with the terms of
licences issued by the Copyright Licensing Agency. Enquiries concerning
reproduction outside those terms and in other countries should be sent to
the Rights Department, Oxford University Press, at the address above.

This book is sold subject to the condition that it shall not,
by way of trade or otherwise, be lent, re-sold, hired out, or otherwise
circulated without the publisher's prior consent in any form of binding
or cover other than that in which it is published and without a similar
condition including this condition being imposed
on the subsequent purchaser.

A catalogue record for this book is available from the British Library

Library of Congress Cataloging in Publication Data
(Data available).

ISBN 0 19 856434 1

Typeset by Cambrian Typesetters, Frimley, Surrey
Printed in Great Britain
by J. W. Arrowsmith Ltd, Bristol

Preface

This book describes the methods developed and/or applied in the 'Cellular metabolism and histochemistry' research group of the Laboratory of Cell Biology and Histology of the University of Amsterdam. The authors are grateful to all the members of this group and to the students and visitors who have participated in the research during their stay in the laboratory for their invaluable contributions. During the process of writing this book the members of the group were (in alphabetical order): Miss Klazina S. Bosch, Prof. Dr Jan James, Dr Geertruida N. Jonges, Dr Arnold Kooij, Mr Frans Marx, Dr Jacques P. M. Schellens, Miss Ilse M. C. Vogels, and Miss Helena Vreeling-Sinderalova. We wish to thank Miss Els M. Tjong Joe Wai for the careful preparation of the manuscript and Mr Jan Peeterse for designing the figures.

Amsterdam   C. J. F. V. N.
April 1992   W. M. F.

Safety

Attention to safety aspects is an integral part of all laboratory procedures and both the Health and Safety at Work Act 1974 and the Control of Substances Hazardous to Health Regulations 1988 impose legal requirements on those persons planning or carrying out such procedures.

In this and other handbooks every effort has been made to ensure that the recipes, formulae, and practical procedures are accurate and safe. However, it remains the responsibility of the reader to ensure that any procedures which are followed are carried out in a safe and competent manner and in compliance with the requirements under the COSHH regulations. Any specific safety instructions relating to items of laboratory equipment must be observed.

Contents

1 Introduction	1
2 Theoretical considerations	3
2.1 Enzyme mechanisms of action and kinetic parameters	3
2.2 Criteria for the validity of enzyme histochemical techniques	9
2.3 Enzyme kinetics *in situ* as demonstrated histochemically	13
2.4 Qualitative versus quantitative enzyme histochemistry	14
2.5 Quantitative histochemistry based on kinetic measurements versus end-point measurements	16
2.6 Principles of enzyme histochemical methods	19
2.7 Unit of comparison of biochemical and quantitative histochemical data	26
2.8 Metabolic fluxes	29
2.9 Immunocytochemical detection of enzyme molecules and detection of mRNAs coding for enzymes by *in situ* hybridization versus *in situ* localization of enzyme reactions	30
3 Preparation techniques	36
3.1 Tissue and cell preparations	36
3.2 Preparation of sections	39
3.3 Cell preparations	43
3.4 Storage of cell and tissue preparations	45
4 Enzyme histochemical methods	48
4.1 Aqueous media	48
4.2 Media containing polyvinyl alcohol	48
4.3 Semipermeable membrane technique	50
4.4 Enzyme histochemical procedures	52
5 Methods developed for clinical and experimental pathology	102
5.1 Oxygen sensitivity test for the diagnosis of malignancy	102
5.2 'Nothing' dehydrogenase reaction for the detection of necrosis by ischaemia	103

5.3 Lysosomal membrane fragility test 104
5.4 Unfixed and undecalcified bone sections 105
5.5 Detection of glucose-6-phosphate dehydrogenase deficiency in erythrocytes 106

Index 111

1 Introduction

The detection of (macro)molecular structures and sequences in tissues and cells for cell biological and diagnostic purposes has been greatly enhanced by histochemistry and cytochemistry. Methodological approaches such as affinity histochemistry, including immunohistochemistry, *in situ* hybridization, and lectin histochemistry, have been and still are of the utmost importance in life sciences. Catalytic histochemistry or enzyme histochemistry was introduced by Gomori as early as 1939 (see Gomori 1952) and has thus existed for much longer than affinity histochemistry, but it never acquired a leading role in cell biology and pathology, despite the fact that many enzymes have a key regulatory role in various cellular processes. Reasons for the slow progress and lack of interest have undoubtedly been the limited number of enzymes whose activity could be demonstrated *in situ* and the fact that most enzymes which could be demonstrated were not key regulatory enzymes and/or had uncertain (patho)physiological roles. Furthermore, preservation of morphology and immobilization of enzyme molecules in tissues was generally achieved by chemical fixation although most enzymes are extremely vulnerable to fixation (Chayen *et al*. 1973; Chalmers and Edgerton 1989; Hopwood 1991). Another reason for the limited application of catalytic histochemistry in cell biological and pathological research has been the lack of consensus about adequate application of histochemical methods for *in situ* demonstration of enzyme activity. Relatively many methods were applied without proper testing of essential conditions which have to be fulfilled before a method may be considered specific, reproducible, precise, and valid (Stoward 1980; Fahimi 1980). Only a limited number of handbooks have been published that describe the available methods in a comprehensive and critical way. These handbooks (Chayen *et al*. 1973; Lojda *et al*. 1979) are still valuable but since their publication many new methods and approaches have been described, especially for the demonstration of the activity of key enzymes. Moreover, the specificity, precision, reproducibility, and validity of enzyme histochemical techniques have been greatly improved, enabling quantitative analysis of enzyme reactions *in situ* by cytophotometry, flow cytometry, or image analysis (for review, see Van Noorden and Butcher 1991). The quantitative approach can now be used to analyse kinetic parameters of enzymes in their cellular environment and their changes due to regulation mechanisms or under pathological conditions.

The present handbook contains a selection of the most important techniques which are currently available for light microscopy. Methods specific for electron microscopical enzyme histochemistry are not described here because it is a field of its own and requires different criteria. For extensive reviews of these electron microscopical methods, see Borgers and Verheyen (1985) and Wohlrab and Gossrau (1992); for a laboratory manual and practical approach similar to the present book, see Van Noorden and Hulstaert (1991).

The methods in this book (described in a manner which can be followed relatively easily) have been selected because they give reliable and specific results. Special attention has been given to control reactions. For a complete review of all enzyme histochemical and cytochemical methods available at present, the reader is referred to Stoward and Pearse (1991).

References

Borgers, M. and Verheyen, A. (1985). Enzyme cytochemistry. *International Review of Cytology*, **95**, 163–227.

Chalmers, G. R. and Edgerton, V. R. (1989). Marked and variable inhibition by chemical fixation of cytochrome oxidase and succinate dehydrogenase in single motoneurons. *Journal of Histochemistry and Cytochemistry*, **37**, 899–901.

Chayen, J., Bitensky, L., and Butcher, R. G. (1973). *Practical Histochemistry*. Wiley, London.

Fahimi, H. D. (1980). Qualitative cytological criteria for the validation of enzyme histochemical techniques. In *Trends in Enzyme Histochemistry and Cytochemistry* (ed. D. Evered and M. O'Connor), pp. 33–51. Excerpta Medica, Amsterdam.

Gomori, G. (1952). *Microscopic Histochemistry. Principles and Practices*. University of Chicago Press, Chicago.

Hopwood, D. (1991). Fixation of tissue for histochemistry. In *Histochemical and Immunocytochemical Techniques. Applications to Pharmacology and Toxicology* (ed. P. H. Bach and J. R. J. Baker), pp. 147–65. Chapman & Hall, London.

Lojda, Z., Gossrau, R., and Schiebler, T. H. (1979). *Enzyme Histochemical Methods. A Laboratory Manual*. Springer Verlag, New York.

Stoward, P. J. (1980). Criteria for the validation of quantitative histochemical enzyme techniques. In *Trends in Enzyme Histochemistry and Cytochemistry* (ed. D. Evered and M. O'Connor), pp. 11–31. Excerpta Medica, Amsterdam.

Stoward, P. J. and Pearse, A. G. E. (1991). *Histochemistry. Theoretical and Applied*, Vol. 3, 4th edn. Churchill Livingstone, Edinburgh.

Van Noorden, C. J. F. and Butcher, R. G. (1991). Quantitative enzyme histochemistry. In *Histochemistry. Theoretical and Applied*, Vol. 3, 4th edn. (ed. P. J. Stoward and A. G. E. Pearse), pp. 355–432. Churchill Livingstone, Edinburgh.

Van Noorden, C. J. F. and Hulstaert, C. E. (1991). Electron microscopical enzyme histochemistry. In *Electron Microscopy of Tissues, Cells, and Organelles* (ed. J. R. Harris), pp. 125–49. Oxford University Press, Oxford.

Wohlrab, F. and Gossrau, R. (1992). *Katalytische Enzymhistochemie. Grundlagen und Methoden fuer die Elektronenmikroskopie*. Gustav Fischer Verlag, Jena.

2 Theoretical considerations

2.1 Enzyme mechanisms of action and kinetic parameters

Biochemical and histochemical enzyme assays are basically similar except that, in the latter, activity is linked topologically to cell or tissue structures. This direct link between localization and function has implications with respect to methodology and also with regard to the activity of enzymes (see also Section 2.3), but the definitions remain the same. These definitions will be explained here briefly. For an extensive review see Dixon and Webb (1979).

An enzyme is a macromolecule, often a (glyco)protein that catalyses the conversion of (a) more or less specific substrate(s) to its product(s). Normally, conversion of the substrate(s) also occurs in the absence of the enzyme but the enzyme is able to speed up the process greatly (for example, the spontaneous conversion of oxygen radicals into hydrogen peroxide and oxygen is 10 000 times slower than that catalysed by superoxide dismutase). The type of conversion and the type of substrate that is converted defines the enzyme involved. On this basis all known enzymes are classified into six categories (Table 2.1). In this classification each enzyme has its own number made up by four digits. The first digit indicates to which of the six classes the enzyme belongs; the second defines the subclass, the third the sub-subclass, and the fourth is the ordinal number of the enzyme. Table 2.1 lists all enzymes for which a histochemical method is described in this book. For an overview of the nomenclature of enzymes, see Dixon and Webb (1979).

Table 2.1. *Enzymes for which a histochemical method is described in Chapter 4. Complete references are provided at the end of Chapter 4*

The page(s) on which each method is found is given in the index under the respective enzyme name.

Enzyme with EC number	Method	Reference
1. Oxidoreductases		
Glycerol-3-phosphate dehydrogenase (NAD$^+$) (1.1.1.8)	Tetrazolium	Stuart and Simpson 1970

Table 2.1. Continued.

Enzyme with EC number	Method	Reference
UDP glucose dehydrogenase (1.1.1.22)	Tetrazolium	McGarry and Gahan 1985
Lactate dehydrogenase (1.1.1.27)	Tetrazolium	Van Noorden and Vogels 1989b
3-Hydroxybutyrate dehydrogenase (1.1.1.30)	Tetrazolium	Rieder 1981
3-Hydroxyacyl CoA dehydrogenase (1.1.1.35)	Tetrazolium	Chambers et al. 1982
Malate dehydrogenase (NAD$^+$) (1.1.1.37)	Tetrazolium	Wimmer and Pette 1979
Malate dehydrogenase (NADP$^+$) (1.1.1.40)	Tetrazolium	Rieder et al. 1978
Isocitrate dehydrogenase (NAD$^+$) (1.1.1.41)	Tetrazolium	Kugler and Vogel 1991
Isocitrate dehydrogenase (NADP$^+$) (1.1.1.42)	Tetrazolium	Kugler and Vogel 1991
Phosphogluconate dehydrogenase (1.1.1.44)	Tetrazolium	Jonges and Van Noorden 1989
Glucose-6-phosphate dehydrogenase (1.1.1.49)	Tetrazolium	Van Noorden 1984
20α-Hydroxysteroid dehydrogenase (1.1.1.62)	Tetrazolium	Robertson et al. 1982
3β-Hydroxy-Δ^5-steroid dehydrogenase (1.1.1.145)	Tetrazolium	Robertson 1979; Gordon and Robertson 1986
Xanthine oxidoreductase (xanthine: NAD$^+$, 1.1.1.204 and xanthine:O$_2$, 1.2.1.37)	Tetrazolium	Kooij et al. 1991
α-Hydroxyacid oxidase (1.1.3.15)	Cerium	Angermueller et al. 1986; Frederiks et al. 1992a
Glycerol-3-phosphate dehydrogenase (1.1.99.5)	Tetrazolium	Martin et al. 1985; Kugler 1991
Aldehyde dehydrogenase (1.2.1.3)	Tetrazolium	Chieco et al. 1986
Benzaldehyde dehydrogenase (1.2.1.7)	Tetrazolium	Chieco et al. 1986
Glyceraldehyde-3-phosphate dehydrogenase (1.2.1.12)	Tetrazolium	De Schepper et al. 1985
Succinate-semialdehyde dehydrogenase (1.2.1.24)	Tetrazolium	Ritter 1973
Succinate dehydrogenase (1.3.99.1)	Tetrazolium	Butcher 1970; Van Noorden and Vogels 1989b
Glutamate dehydrogenase (1.4.1.2)	Tetrazolium	Wimmer and Pette 1979; Kugler, 1990a
D-Amino acid oxidase (1.4.3.3)	Cerium	Frederiks et al. 1992a
Monoamine oxidase (1.4.3.4)	Tetrazolium	Frederiks and Marx 1985
NADPH-ferrohaemoprotein reductase (1.6.2.4)	Tetrazolium	Van Noorden and Butcher 1986
NAD(P)H dehydrogenase (1.6.99.2)	Tetrazolium	Straatsburg et al. 1989
Urate oxidase (1.7.3.3)	Cerium	Angermueller 1989; Frederiks et al. 1992a
Cytochrome c oxidase (1.9.3.1)	Diaminobenzidine	Angermueller and Fahimi 1981; Kugler et al. 1988b; Hiraoka et al. 1986

Peroxidase (1.11.1.7)	Diaminobenzidine	Angermueller and Fahimi 1981; Hayhoe and Quaglino 1988
Catalase (1.11.1.6)	Diaminobenzidine	Angermueller and Fahimi 1981
Monophenol monoxygenase (1.14.18.1)	Natural chromophore	Whittaker 1981
Cytochrome P450	Natural chromophore	Watanabe et al. 1989

2. Transferases

Ornithine carbamoyltransferase (2.1.3.3)	Metal salt	Holzgreve et al. 1985
γ-Glutamyltransferase (2.3.2.2)	Diazonium	Gossrau 1985
Glycogen phosphorylase (2.4.1.1)	Synthesis reaction	Frederiks et al. 1987b
Purine nucleoside phosphorylase (2.4.2.1)	Metal salt / Tetrazolium	Van Reempts et al. 1988 / Frederiks et al. 1992b
Aspartate aminotransferase (2.6.1.1)	Metal salt	Papadimitriou and Van Duijn 1970
γ-Aminobutyric acid transaminase (2.6.1.19)	Tetrazolium	Kugler and Baier 1990
Hexokinase (2.7.1.1)	Tetrazolium	Kugler 1990b
Phosphofructokinase (2.7.1.11)	Tetrazolium	Frederiks et al. 1991
NAD^+-kinase (2.7.1.23)	Tetrazolium	Macha et al. 1975
Pyruvate kinase (2.7.1.40)	Tetrazolium	Klimek et al. 1988
Creatine kinase (2.7.3.2)	Tetrazolium	Frederiks et al. 1988b

3. Hydrolases

Non-specific esterases (3.1.1.–)	Diazonium	Kaplow et al. 1976; Lojda et al. 1979
Acetylcholinesterase (3.1.1.7)	Thiocholine	Kugler 1987
Alkaline phosphatase (3.1.3.1)	Indoxyl-tetrazolium	Van Noorden and Jonges 1987
Acid phosphatase (3.1.3.2)	Diazonium	Frederiks et al. 1987c; Van Noorden et al. 1989a
5′-Nucleotidase (3.1.3.5)	Metal salt	Frederiks and Marx 1988
Glucose-6-phosphatase (3.1.3.9)	Cerium	Jonges et al. 1990
Arylsulfatase (3.1.6.1)	Metal salt	Hopsu-Havu et al. 1967
Acid α-glucosidase (3.2.1.20)	Diazonium	Gutschmidt et al. 1979
α-Galactosidase (3.2.1.22)	Diazonium	Lojda et al. 1979
β-Galactosidase (3.2.1.23)	Indigogenic	Lund-Hansen et al. 1984
Lactase (3.2.1.23)	Diazonium	Lojda et al. 1979
α-Mannosidase (3.2.1.24)	Diazonium	Lojda et al. 1979
N-Acetyl-β-glucosaminidase (3.2.1.30)	Diazonium	Lojda et al. 1979; Robertson 1980
β-Glucuronidase (3.2.1.31)	Diazonium	Schofield et al. 1983
Aminopeptidase M (3.4.11.2)	Diazonium	Wachsmuth and Donner 1976
Aminopeptidase A (3.4.11.7)	Diazonium	Kugler 1982a
Dipeptidyl peptidase II (3.4.14.2)	Diazonium	Gossrau and Lojda 1980
Dipeptidyl peptidase IV (3.4.14.5)	Diazonium	Gossrau 1985
Elastase (3.4.21.11)	Fluorescence	Rudolphus et al. 1992
Cathepsin B (3.4.22.1)	Diazonium / Fluorescence	Van Noorden et al. 1989b / Van Noorden et al. 1987
Guanine deaminase (3.5.4.3)	Tetrazolium	Ito et al. 1988
ATPases (3.6.1.3)	Metal salt	Firth 1987

4. Lyases

Ornithine decarboxylase (4.1.1.17)	Metal salt	Dodds et al. 1990
Fructose-biphosphate aldolase (4.1.2.13)	Tetrazolium	Lojda et al. 1979
Adenylate cyclase (4.6.1.1)	Metal salt	Mayer et al. 1985

Table 2.1. *Continued.*

Enzyme with EC number	Method	Reference
5. Isomerases		
Glucose-6-phosphate isomerase (5.3.1.9)	Tetrazolium	De Schepper *et al.* 1985
Phosphoglucomutase (5.4.2.2)	Tetrazolium	De Vries and Meijer 1976
6. Ligases (synthetases)		

The activity of an enzyme is defined by its kinetics. These kinetics can be, but are not necessarily, very complicated. In histochemistry, the zero-order velocity of enzymes is usually determined; this is nearly equal to the maximum velocity of an enzyme (V_{max}) because relatively high substrate concentrations are used. Enzyme activity depends on the substrate concentration (Fig. 2.1) up to a point at which increasing the concentration of the substrate no longer results in an increase in the velocity of the enzyme reaction. This is when the reaction is said to be of zero-order, i.e. that it is independent of the substrate concentration. The activity then approaches V_{max}, but the V_{max} can never be reached because it is the velocity of the enzyme at an indefinitely high substrate concentration. However, V_{max} is one of the important kinetic parameters which determine the activity profile of an enzyme. It can be calculated by analysing the enzyme reaction at different substrate concentrations (Figs. 2.1 and 2.2a) of which a substantial number should be below the K_M value. The K_M value is the substrate concentration giving an enzyme reaction velocity of $\frac{1}{2}V_{max}$ (Fig. 2.1). Figs. 2.1 and 2.2 show

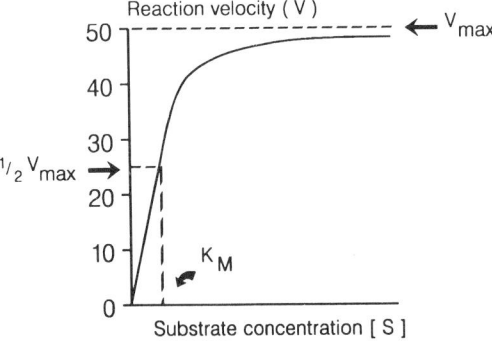

Fig. 2.1. A plot of the reaction velocity, V, as a function of the substrate concentration, [S], for an enzyme that obeys Michaelis–Menten kinetics; V_{max} is the maximal velocity, a parameter for the amount of enzyme, and K_M is the Michaelis constant, which reflects the affinity of an enzyme for its substrate.

Theoretical considerations 7

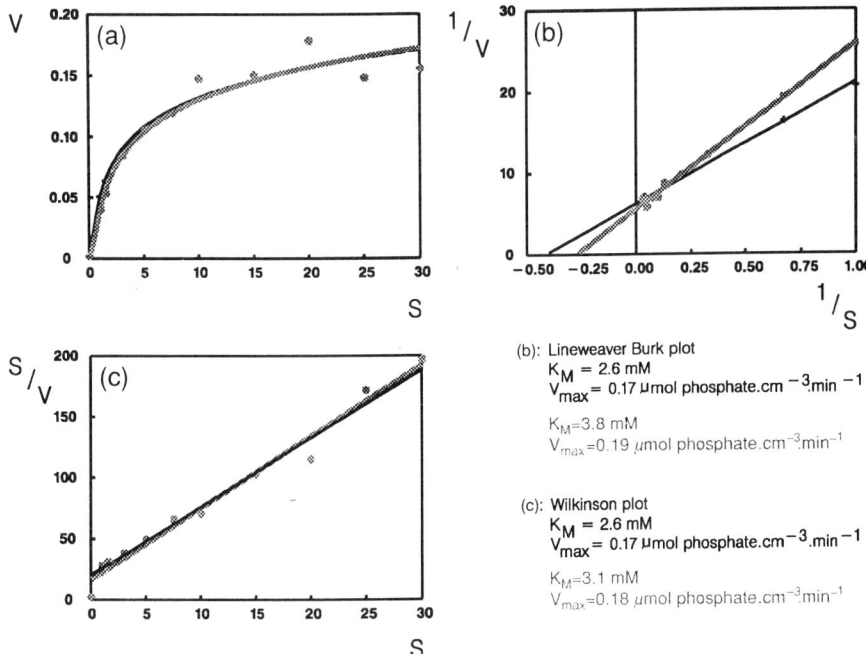

Fig. 2.2. Determination of kinetic parameters of an enzyme (glucose-6-phosphatase) reaction by quantitative histochemistry. (a) Plot demonstrating the relation between the substrate concentration (S, in mM) and the reaction velocity (V, in μmol phosphate liberated per cm^3 wet weight of tissue per minute). (b) Mathematical conversion of data from (a) into a Lineweaver–Burk plot ($1/S$ versus $1/V$). (c) Mathematical conversion of data from (a) into a Wilkinson plot (S versus S/V). Black line, experimental data; grey line, similar data except that the values for V measured at the two lowest substrate concentrations were decreased by 20 per cent (see text). These small errors have a large impact on (b) but not on (c).

that both kinetic parameters can be calculated from a series of reactions at different substrate concentrations. Listing 2 gives the procedure for histochemical assays.

Several mathematical conversions of the relationship between substrate concentrations and reaction velocity exist. The Wilkinson plot (Wilkinson 1961) is considered to be the best because every point in this relationship is statistically equally important (Fig. 2.2). The Lineweaver–Burk plot is used much more frequently for this conversion but is statistically less valid. Some points (particularly those relating to low substrate concentrations and thus low reaction velocities) have a larger impact on the conversion than others in this plot (Fig. 2.2). These measuring points are more inaccurate, because in general, non-specific control reactions (background reactions) have to be subtracted from test reactions (Fig. 2.3a) in order to obtain a specific 'test minus control' reaction. Control reactions are usually rather high in histochemical assays due to the presence in sections of many reactive

8 *Enzyme histochemistry: a laboratory manual of current methods*

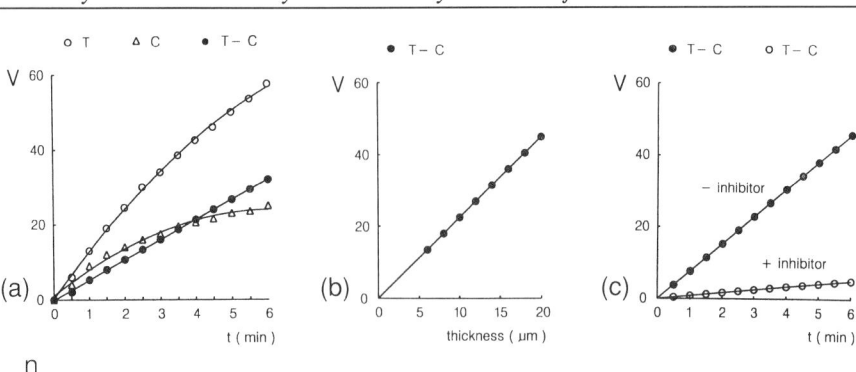
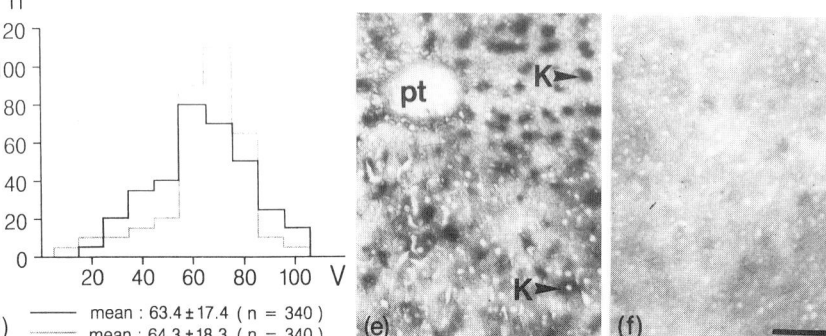

Fig. 2.3. Examples of validity tests for (quantitative) enzyme histochemical methods. (a) Relationships between the formation of final reaction product (*V*) as parameter for the velocity of an enzyme reaction and incubation time as measured kinetically in an unfixed cryostat section (see Section 2.5); T, test reaction in the presence of substrate; C, control reaction in the absence of substrate; T − C, 'test minus control' reaction. (b) Relationship between the amount of final reaction product formed (*V*) after a set period of incubation and section thickness for the specific 'test minus control' reaction (T − C) of an enzyme *in situ*. (c) Relationships for the specific 'test minus control' reaction of an enzyme in unfixed cryostat sections between the formation of final reaction product (*V*) and incubation time as measured kinetically in the presence or absence of a specific inhibitor. (d) Reproducibility of an enzyme reaction *in situ* as measured in unfixed cell preparations incubated on two different days. The histograms show the cytophotometric analysis of final reaction product precipitated in 340 individual cells in each assay. The histograms are sufficiently similar to conclude that the method is reproducible. (e and f) Demonstration of high (e) and low (f) precision of an enzyme histochemical method. The only difference between (e) and (f) is that the concentration of tetranitro BT (1 mM instead of 5 mM) used in (f) is too low for the demonstration of glucose-6-phosphate dehydrogenase activity in unfixed cryostat sections of rat liver (see Method 33). Due to the suboptimal medium in (f), the high activity in Kupffer cells is barely visible. K, Kupffer cell; pt, portal tract. Bar, 50 μm.

molecules that may interfere with the enzyme reaction (Fig. 2.3a; see also Section 2.3). As can be concluded from Fig. 2.2, data on the glucose-6-phosphatase reaction in periportal zones of rat liver yield K_M values of 2.6 and 2.7 mM and V_{max} values of 157 and 162 μmol $PO_4^{3-} \cdot cm^{-3} \cdot min^{-1}$ when calculated from a Wilkinson plot or a Lineweaver–Burk plot, respectively. These values are virtually the same, but if, for example the values measured at substrate concentrations of 1 and 1.5 mM had been 20 per cent lower due to experimental error (i.e. 48 and 52 instead of 58 and 62, respectively), then the K_M and V_{max} values calculated from the Wilkinson plot would be 2.7 and 172. If the Lineweaver–Burk plot was used, the K_M and V_{max} would be 3.8 and 187. So for histochemical assays it is recommended that the K_M and V_{max} values are calculated using Wilkinson plots.

It should be realized that the K_M and V_{max} values depend on assay conditions such as pH and temperature of the medium, and presence of activators. Moreover, as will be shown in Section 2.3, the protein concentration in a solution containing the enzyme, and thus the environment of the enzyme, is particularly important for the enzyme's kinetic properties.

Many enzymes have more than one substrate, e.g. glucose-6-phosphate dehydrogenase simultaneously oxidizes glucose-6-phosphate into 6-phosphoglucono-δ-lacton and reduces NADP into NADPH in a redox reaction. For these enzymes, the determination of the kinetics is more laborious but in principle not more complicated. For both substrates the K_M and V_{max} have to be calculated; this is done by keeping one substrate concentration in the reaction medium constant while varying the other and vice versa.

Another important parameter of an enzyme is its inhibitor constant (K_i). By varying the substrate concentration at two, but preferably three or four, inhibitor concentrations in the medium, it is possible to determine not only the sensitivity of an enzyme for the inhibitor (expressed as K_i) but also the type of inhibition (competitive, non-competitive, allosteric etc.). This is particularly important with respect to the specificity of the reaction. For an overview of all enzymes with their known inhibitors and the inhibitor constants, see Zollner (1989).

Recently, it has become possible to measure substrate concentrations in tissue sections or complete cells (Mueller-Klieser *et al.* 1988; see Section 2.8). This is particularly important because it enables the calculation of substrate fluxes through metabolic pathways which are important parameters for cellular metabolic activity (Newsholme *et al.* 1980). For the calculation of fluxes, the V_{max}, K_M, and actual substrate concentration (s) in a tissue need to be known; the substrate flux (v) can be calculated from $v = (V_{max} \cdot s)/(K_M + s)$ (Dixon and Webb 1979).

2.2 Criteria for the validity of enzyme histochemical techniques

Fahimi (1980) and Stoward (1980) formulated a set of criteria which have to be fulfilled before an enzyme histochemical method may be said to be precise, specific, reproducible, and valid. A method is said to be *precise* when the final reaction product is precipitated at the true *in vivo* subcellular site of the enzyme. A method is said to be *specific* when the final reaction product is generated only by the enzyme under study. Control reactions to test the specificity are essential for each enzyme histochemical method and therefore emphasis is given here to control reactions for all methods described. A method is said to be *reproducible* when it gives similar results irrespective of the place and time of its performance. A *valid* method gives rise to formation of an amount of final reaction product proportional to the *in situ* activity of the enzyme. For example, the latter requirement is usually not met in electron microscopy because tissue preservation by fixation leads to in-activation of at least some of the enzyme molecules and inactivation may not occur to the same extent in different cellular compartments. For light microscopy, this criterion is met more easily because unfixed tissue can be used.

In more practical terms, a number of specific criteria must be met before an enzyme histochemical method may yield a correct localization of enzyme activity in tissue sections or cell preparations. The most important of these specific criteria are:

- The colour reagent or indicator must be water-soluble and preferably colourless; it should be converted by the specific enzyme reaction into a water-insoluble final reaction product that is coloured (or electron-dense in the case of electron microscopy).
- There should be no diffusion of enzyme molecules, intermediate reaction products, or the final coloured reaction product from sections or cells into the incubation medium, because this yields erroneous information about the localization of the enzyme under study.
- Pretreatment of tissues or cells (such as fixation) should not cause changes in the activity or distribution pattern of the enzyme activity.
- All compounds necessary for the enzyme reaction should be able to diffuse sufficiently quickly to the active sites of the enzyme to ensure that it is the activity of the enzyme, and not the diffusion processes, that determines reaction velocity and formation of the final coloured reaction product.
- The formation of the final reaction product due to enzyme activity should be proportional to time and to the activity of the enzyme.

Whether these criteria can be met depends on both the enzyme incubation step(s) and tissue pretreatment before incubation. Either fixed or unfixed tissues can be used as well as mounted or unmounted tissues. Fixation and mounting in resins, when properly performed, enable the correct subcellular localization of the final reaction product to be retained (high precision), but both fixation and mounting are detrimental for most enzymes. On the other hand, the use of unfixed and unmounted sections for enzyme histochemistry can result in poor morphology if precautions are not taken during the incubation step(s). However, a number of relatively easily applicable measures, such as the use of a semipermeable membrane between an unfixed tissue section and gelled incubation medium or the addition of a tissue stabilizer to the medium such as polyvinyl alcohol, circumvent a number of these drawbacks (see Sections 2.6.10, 4.2, and 4.3). Therefore, the most of the methods described here are performed on cryostat sections of unfixed tissues with the use of polyvinyl alcohol or semipermeable membranes. The great advantage is that pretreatment is unnecessary and all enzyme activity present in the tissue remains *in situ*. For a recent review of the use of polyvinyl alcohol and semipermeable membrane techniques for enzyme histochemical purposes, see Van Noorden and Vogels (1989).

On the other hand, certain pretreatment methods are available that ensure maximum retention of morphology without inactivating all enzyme activity. The principle of these methods is to perform fixation as mildly as possible, and to mount the tissue or cell smear in water-miscible resins that polymerize at low temperature. However, despite the fact that precise localization at the subcellular level can be obtained when all procedures are carried out carefully, reproducibility and validity of the method may not be achieved because inactivation of enzymes may vary distinctly at different sites within a tissue or cell.

Testing of the criteria (see Listing 1) can essentially be performed in two ways: firstly, by qualitative inspection of the tissue sections after incubation (for example to determine the precision of the method) and secondly, by using cytophotometric means for the quantification of the amount of final reaction product precipitated in a tissue specimen after incubation. All incubation steps have to be performed under rigidly constant conditions. These conditions are essentially the same as for biochemical assays but in histochemistry these rigid conditions are still not applied as uniformly as they should be.

Listing 1. *Criteria for reliability of enzyme histochemical reactions (Stoward 1980)*

Precision
1. Sections retain their morphology and look 'clean'.
2. Specific final reaction product is confined to certain subcellular sites known to contain the enzyme.
3. Specific final reaction product does not diffuse and/or precipitate at other (sub)cellular sites.

Specificity
4. No specific final reaction product is formed in control sections or preparations incubated without substrate.

5. The reaction conditions that give rise to maximum rates of formation of specific final reaction product *in situ* are similar to those in *in vitro* systems.
6. Inhibitors and activators exert their specific effects on the formation of final reaction product in a similar way to *in vitro* systems.
7. Potentially interfering enzyme systems have been either suppressed or shown to be absent, or can be distinguished from the enzyme under study.

Reproducibility
8. The mean values of measurable parameters (for example absorbance or fluorescence of the final reaction product) do not vary significantly in repeated experiments.
9. Individual measurements of these parameters within a preparation form a statistically unimodal population.

Validity
10. No enzyme is lost from its (sub)cellular site during procedures required for the visualization of its activity.
11. Specific final reaction product generated in an enzymic reaction has the expected chemical composition.
12. Specific final reaction product generated in an enzymic reaction can be identified in cells or tissues.
13. There is a stoichiometric relationship between the amount of final reaction product precipitated and the amount of primary reaction product formed by the enzyme.
14. The mean absorbance or fluorescence of the specific final reaction product is proportional to its concentration in the cell or section.
15. The amount of final reaction product formed specifically is proportional with enzyme activity (or with section thickness).
16. The amount of final reaction product formed specifically is proportional with incubation time.
17. On extrapolated plots of absorbance or fluorescence against incubation time, the absorbance or fluorescence of the specific final reaction product should be zero or at worst small but constant at time zero.
18. The rate of formation of final reaction product is a function of the substrate concentration in the incubation medium.

Quantification requires testing with criteria 13–18 of Listing 1 above to determine whether the specific ('test minus control') reaction is proportional to time (Fig. 2.3a) or to the amount of enzyme (or to section thickness; Fig. 2.3b). It also enables estimation of specificity by following both test reactions (in the presence of substrate) and control reactions (in the absence of substrate and/or in the presence of a specific inhibitor) (see Section 2.5). The specific 'test minus control' reaction should be proportional to time (Fig. 2.3a) and inhibitable by specific inhibitor(s) (Fig. 2.3c). Reproducibility is tested by repeating the reaction within one assay (intra-assay variation) by incubating a number of sections at the same time, whereas performing the same experiment on different days enables the estimation of inter-assay variation (Fig. 2.3d). An example of high and low precision of localization is shown in Fig. 2.3e and f; in the latter, precision is lost due to the use of a suboptimal incubation medium. See Van Duijn (1991) and Van Noorden and Butcher (1991) for extensive discussions of validity tests of histochemical assays.

2.3 Enzyme kinetics *in situ* as demonstrated histochemically

After satisfying the validity tests discussed in Section 2.2, the histochemical reaction can be used for the analysis of kinetic parameters of an enzyme under the conditions of a histochemical assay (see Listing 2). These conditions are significantly different from those of biochemical assays. Firstly, the enzyme molecules are present in their *in vivo* cellular environment when analysed *in situ* by histochemical means. The cytoplasmic matrix is a very concentrated protein solution of 20–40 per cent by weight. This has implications with respect to the water phase in cells because a significant part of the water present in cells is adsorbed to proteins and is osmotically inactive; it behaves differently in its solvent properties. Furthermore, the high cellular protein concentration leads to self-association or hetero-association which means that proteins form dimers or polymers or bind to structural elements such as the cytoskeleton (for review, see Van Noorden and Gossrau 1991). In biochemical assays, enzymes are usually (partially) purified and, therefore, the protein concentration in the reaction media is very low (0.1 per cent or less). Although these assays reveal the kinetic properties of the enzyme *per se*, these may not reveal how the enzyme works *in vivo* (Dixon and Webb 1979; Groen *et al.* 1982). Secondly, enzymes may be present in a cellular compartment such as lysosomes or as part of membrane systems such as the plasma membrane which may also affect the working mechanism of an enzyme. These properties are usually lost after the enzyme has been purified. Therefore, it is not surprising that the kinetic properties of enzymes as analysed histochemically *in situ* can be entirely different from those analysed biochemically *in vitro*. Table 2.2 gives an example of this phenomenon for hexokinase (Lawrence *et al.* 1989).

14   *Enzyme histochemistry: a laboratory manual of current methods*

Listing 2. *Determination of kinetic parameters of enzymes in situ as analysed histochemically*

1. Use unfixed serial cryostat sections of one thickness (for instance, 8 or 10 μm) or unfixed cell preparations.
2. Prepare fresh incubation media with different concentrations of substrate (see, for example, Fig. 2.2a) according to the methods in Chapter 4.
3. Incubate at least three sections for each substrate concentration during a set period of time at constant temperature (preferably 37 °C).
4. Rinse sections thoroughly and mount in glycerine–gelatin.
5. Measure absorbance at the absorption maximum or at the isobestic wavelength of the final reaction product (see Van Noorden and Butcher 1991) with a scanning and integrating cytophotometer or image analyser in at least five selected areas per section.
6. Calculate mean absorbance values for each substrate concentration and subtract the correct control value (usually the reaction in the absence of substrate (see Sections 2.1 and 2.5).
7. Convert data to absolute units of enzyme activity when possible (see Method 3).
8. Plot data on a Wilkinson plot (see Fig. 2.2) and calculate the best fitting line $y = ax + b$; from this equation calculate $K_M = b/a$ and $V_{max} = 1/a$.

Table 2.2. K_M *values (in mM) for rat hexokinase type I in cerebellum as determined by histochemical and biochemical assays (after Lawrence et al. 1989)*

Assay	Sample preparation/pretreatment	K_M (mM)
Histochemical (+15% PVA)	Unfixed sections	0.13
(−PVA)	Acetone fixed sections	0.48
Biochemical	Purified enzyme	0.05
	5 per cent (w/v) homogenate	0.06
	10 per cent (w/v) homogenate	0.12
	25 per cent (w/v) homogenate	0.28

Abbreviation: PVA, polyvinyl alcohol

2.4 Qualitative versus quantitative enzyme histochemistry

One of the main purposes of this book is to demonstrate that enzyme histochemistry has now reached the level where in principle every aspect of cellular metabolism can be analysed as it occurs *in situ* (probably reflecting the situation *in vivo*). Nevertheless, more fundamental research is needed to extend the range of techniques available to the histochemist. As has been discussed in the previous sections, kinetic parameters of enzymes, the effect of inhibitors, and substrate fluxes through metabolic pathways can be determined using quantitative histochemistry. The instrumentation needed for this quantitative approach is more complex than the spectrophotometer or fluorimeter needed for biochemical assays. This is because a heterogeneously distributed final reaction product is measured which may lead to serious distributional errors (see James (1983) and Van Noorden (1989) for reviews of instrumental aspects of measurement). Therefore, either a scanning and integrating cytophotometer or image analysing system is used when measuring a coloured final reaction product (for reviews, see Lawrence *et al.* 1990; Van Noorden and Butcher 1991) or a static or flow cytofluorimeter when the final reaction product is fluorescent (for review, see Van Noorden 1991). Nowadays, image analysing systems or flow cytofluorimeters are available in most cell biology or pathology laboratories. We recommend quantitative rather than qualitative analysis of the formation of final reaction product because numbers are easier to work with and also give more objective information with respect to parameters such as control reactions, reactions in time etc. as discussed in Sections 2.2 and 2.5.

Considerably more valid methods are available that yield specific light absorbing final reaction products than those that yield fluorescent final reaction products. Image analysis would, therefore, seem to be the instrumental approach for the measurement of *in situ* enzyme reactions for the near future. The main problem with fluorescence is that hardly any enzyme histochemical methods are available that give rise to sufficiently water-insoluble fluorescent final reaction products (Raap 1986). The only exception is the nitrosalicylaldehyde method for proteases (Section 2.6.8). On the other hand, the discovery that coloured final reaction products can quench fluorescence in a stoichiometric way may lead to further developments in this direction. The signal to be quenched can be e.g. autofluorescence generated by glutaraldehyde fixation (Van Noorden *et al.* 1989).

Besides the quantitative approaches, it will remain possible to use qualitative enzyme histochemistry. Depending on the cell biological question to be answered this can be a perfectly valuable procedure. However, we would stress that qualitative methods should also be tested against the criteria given in Section 2.2. For example, the intensity of the coloured final reaction product should indicate the effective activity of the enzyme to be studied. This does not always give the most beautifully stained preparations but the result is meaningful. The reaction should be performed at zero-order kinetics so that the maximum capacity of an enzyme is demonstrated. When these conditions are seriously taken into consideration, the results of qualitative enzyme histochemistry will be scientifically sound. In the past, this often has not been the case, which has contributed to the reputation for unreliability of histochemical tests.

2.5 Quantitative histochemistry based on kinetic measurements versus end-point measurements

Reaction rates of enzymes can be calculated either from end-point measurements after a set incubation time or from kinetic measurements where the formation of the final reaction product is monitored during incubation. Enzyme reactions in tissue sections or cell preparations are most commonly assessed by measuring the amount of final reaction product after it has been precipitated for a set incubation time. Such end-point measurements are only valid when the amount of final reaction product generated is proportional to the incubation time. However, most reactions in sections or cell preparations show non-linearity with time, as has been found by continuous monitoring (Fig. 2.3a) (Van Noorden and Butcher 1991). The high initial reaction rate decreases rapidly as the incubation time increases. This is generally due to endogenous activity which is also found after control incubations. The specific reaction of the enzyme to be studied can only be found after subtraction of the control reaction from the test reaction (Fig. 2.3a). Proper controls are obtained by carrying out the reaction either in the presence of a specific inhibitor of the enzyme or in the absence of substrates (and cofactor); such controls must be performed under the same conditions and measured in similar areas of serial sections to those used for the test reaction. The specific 'test minus control' reaction should be linear with incubation time for at least some minutes as shown in Fig. 2.3a. Test and control reactions are measured in serial sections by incubating the sections on the stage of a microscope attached to a cytophotometer or image analyser. Subtraction of the control reaction from the test reaction yields the linear 'test minus control' reaction that is specific for the enzyme in question (see Method 1).

Method 1. *Procedure for end-point measurements*

1. Use unfixed serial cryostat sections or cell preparations.
2. Prepare fresh incubation media (both for test and control reactions) according to the Methods in Chapter 4 and allow to equilibrate to the incubation temperature.
3. Remove sections from the cryostat cabinet prior to incubation and allow to equilibrate to the incubation temperature for 5–10 min.
4. Pour incubation medium onto sections. Use three to five sections for each incubation condition. Use incubation periods that guarantee a linear 'test

minus control' relationship between the final reaction product and incubation time. When this is not known, incubate for various periods of time and determine the correct period of incubation first.
5. Pour off the medium from the sections and mount in glycerine–gelatin.
6. Measure the absorbance in at least five different areas per section using a scanning and integrating cytophotometer or image analyser.
7. Calculate the means of test and control reactions and convert the arbitrary machine units to mean integrated absorbance values as described in Method 3. When possible, convert these values to absolute units of substrate used per unit volume of wet tissue and per unit time according to Method 3.

The phenomenon of a linear 'test minus control' reaction specific for an enzyme has been found for many enzymes irrespective of the tissue tested (see e.g. a special issue of *The Histochemical Journal*, volume 21, pp. 515–624, 1989). Kinetic measurements can be applied only when the histochemical method allows a one-step reaction such as in reactions based on reduction of a tetrazolium salt (see Section 2.6.5), and simultaneous coupling reactions with the use of a diazonium salt (see Section 2.6.3). Two-step methods based on an incubation step for the enzyme reaction and a subsequent step to convert an intermediate reaction product into a coloured final reaction product (such as post-coupling methods and metal salt capture methods) are not appropriate because the rate of the second step cannot be taken as a parameter for the reaction rate of the enzyme. This also implies that in one-step reactions, a second or third reaction (when needed) has to take place much faster than the enzyme reaction so that the generation of coloured final reaction product is a direct indicator of the enzyme reaction (Van Duijn 1991). Tetrazolium salt and simultaneous coupling reaction methods meet this criterion under optimum conditions. Most of the *in situ* reaction rate studies published so far are on dehydrogenases and reductases using tetrazolium salt methods. Furthermore, simultaneous coupling reactions with diazonium salts for proteinases and reactions for phosphofructokinase and alkaline phosphatase have also been successfully adapted (for review, see Van Noorden and Butcher 1991). Method 2 lists a general procedure for kinetic analysis of enzyme reactions *in situ*.

The equipment needed for quantitative histochemical analysis on either an end-point or a kinetic basis is a cryostat (preferably motor-driven to ensure constant cutting speed and thus constant section thickness) and either a scanning and integrating cytophotometer or an image analyser. At present, the latter system has become the equipment of choice because hardware and appropriate software are now available at a reasonable price, as well as sensitive CCD cameras (James and Tanke 1991). The great advantage of an image analyser over scanning and integrating cytophotometers is the capacity of the former to store in the memory of its computer two-dimensional images of enzyme reactions as they proceed in time in the microscopical field. When a reaction is finished, appropriate areas within the field can be selected from

18 *Enzyme histochemistry: a laboratory manual of current methods*

Method 2. *Procedure for in situ enzyme reaction rate studies*

1. Use unfixed cryostat sections or cell preparations.
2. Keep serial sections in the cryostat cabinet until used.
3. Remove a section from the cabinet prior to incubation and allow to equilibrate to the incubation temperature for 5–10 min. Use this period of time to select an area in the section to be measured which can also be recognized in following sections for measuring control reactions or reactions under different conditions (such as variation of substrate concentrations).
4. At time zero, pour the incubation medium, which has to be at exactly the desired incubation temperature, into a Perspex ring surrounding the section. Relocate the selected area and focus. After some practice, this can be done within 10 s after the medium is poured on top of the section.
5. Make the first reading after 15 s. It lasts one or two seconds. Subtract this 'blank' value from each further reading. Take time point 15 s as zero.
6. Make subsequent measurements at 15 s intervals when the reaction is fast or at 30 s intervals for slower reactions.
7. Analyse the reaction for at least 5 min.
8. Repeat the procedures of steps 1–7 with a consecutive section and a control incubation medium (lacking substrate or containing a specific inhibitor).
9. Repeat both test and control reactions at least three times at constant incubation temperature.
10. Calculate the mean of both test and control reactions and plot these as in Fig. 2.3a.
11. Subtract the control reaction from the test reaction and analyse with regression analysis whether the 'test minus control' reaction is linear with time.
12. Convert the machine units of absorbance to mean integrated absorbance (Method 3) and when possible subsequently to absolute units of substrate converted per unit volume of wet tissue and per unit time (Method 3).

Notes: When applying Perspex rings one needs an objective with long working distance which may be used to focus without a cover glass; particularly when higher magnifications are needed, use a cover glass over the incubation medium.

It is recommended that a spacer is inserted between the slide and cover glass to ensure a certain volume of incubation medium; in this case almost all objectives can be used.

Computer software has been developed for automation of cytophotometric measurements and storage of data which facilitates the whole procedure considerably. The computer immediately produces plots of reaction rates (for review, see Van Noorden and Butcher 1991).

The system developed by Pette and co-workers (Pette and Wimmer 1980) is a photometer equipped with a microscope stage moved by two stepping motors and controlled by a computer; a maximum of 12 different areas can be selected within one section. Once the reaction is started, the increase in absorbance at each position and the duration of the incubation are recorded. One cycle of absorbance measurements of the 12 preselected areas takes no more than 20 s.

these images for analysis of the reaction in time. When the cytophotometer is used, one has to select an area to be measured in an unstained section before the reaction is started. Moreover, only one-dimensional scans of a particular area can be made at best (Van Noorden and Gossrau 1991).

Kinetic measurements are to be preferred when quantitative analysis is performed because the relationship between the formation of the final reaction product and time is accurately monitored in different areas within one section, particularly when image analysis is applied. On the other hand, once it is known that in a certain tissue the 'test minus control' reaction has been thoroughly characterized and is linear with time, end-point measurements are sufficient. The reactions should then be performed during a set incubation time known to give a linear relationship between specific final reaction product formation and time (for example, up to 5 min in Fig. 2.3a). These end-point measurements can be used for the determination of kinetic parameters (see Method 1).

2.6 Principles of enzyme histochemical methods

The most reliable and valid enzyme histochemical methods are discussed here and Table 2.1 summarizes which enzymes can be demonstrated validly with these methods.

2.6.1 Metal salt capture methods

Metal salt capture methods are used to localize the activity of phosphatases, sulfatases, nucleotidases, and cholinesterases. Physiological substrates are usually applied. The general reaction mechanism can be best explained by using alkaline phosphatase as an example: β-glycerophosphate is split at alkaline pH in the presence of Ca^{2+} ions into glycerol and phosphate (primary or splitting reaction). Phosphate is immediately captured by Ca^{2+} ions but calcium phosphate is neither coloured nor electron dense, and therefore a post-treatment is necessary in order to visualize the reaction product. The tissue specimen is incubated in an aqueous solution of cobalt chloride and all calcium phosphate is transformed into pinkish cobalt phosphate, because cobalt phosphate is less water-soluble than calcium phosphate (first transformation reaction). For light microscopy, a second transformation step is necessary in order to visualize cobalt phosphate by incubation of the specimen in a solution of ammonium sulfide. Blackish-brown cobalt sulfide is generated where cobalt phosphate was precipitated. This latter transformation reaction is not necessary for electron microscopy because cobalt is sufficiently electron dense.

Demonstration of acid phosphatase activity with metal salt capture methods is based on the same principle but, as the splitting reaction has to be performed at low pH and as calcium phosphate is rather water-soluble at acid pH, calcium ions are replaced by lead ions in the splitting reaction. On the other hand, lead ions cannot be used at alkaline pH because lead salts precipitate at higher pH values. For electron microscopy, formation of lead phosphate in the splitting reaction is sufficient to obtain a visible product, because lead is electron dense. For light microscopy, lead phosphate is converted into lead sulfide in the visualization reaction.

Precise and reliable localization is sometimes not possible because the precipitation reaction is too slow. The primary reaction product can then diffuse from the enzymic sites before it is captured. This occurs mainly when the local concentration of the capturing reagent is not sufficiently high e.g. due to penetration problems (Van Duijn 1991).

Sulfatase, nucleotidase, and cholinesterase activities are demonstrated in a similar way using substrates specific for the respective enzymes.

2.6.2 Cerium salt capture methods

Cerium salt capture methods are relatively new metal salt capture methods. At present cerium is considered to be superior to lead or cobalt because of the fine granular aspect of cerium salts and fewer problems with non-specific precipitates (Hulstaert *et al.* 1989). Cerium is sufficiently electron dense to be detected by electron microscopy and recent developments enable cerium salts to be used for light microscopy as well. Cerium phosphate can be converted into lead phosphate and subsequently into lead sulfide. However, better results are obtained by converting cerium phosphate using hydrogen peroxide and diaminobenzidine. Diaminobenzidine is oxidized to brown polymerized diaminobenzidine via oxygen radical formation. These radicals are generated by decomposition of cerium perhydroxide which is formed via the conversion of cerium phosphate and hydrogen peroxide into cerium perhydroxide.

Cerium perhydroxide can also be generated by the activity of oxidases in the presence of cerium ions because oxidases produce hydrogen peroxide. These cerium-based methods for oxidases have become very important, both for light and electron microscopical studies of the role of (peroxisomal) oxidases in all kinds of (patho)physiological processes (Angermueller 1989).

2.6.3 Diazonium salt methods

Artificial substrates, containing a 1- or 2-naphthol, naphthol anilid acid (naphthol AS) or naphtholamine group, are used in diazonium salt methods. These methods can be applied to localize the activity of phosphatases, glycosidases, esterases, and proteases. Depending on the enzyme activity to be localized, the substrate contains a group that can be specifically cleaved from the naphthol group in the primary or splitting reaction. The naphthol group can then be coupled with a diazonium salt to produce a coloured azo-dye in the coupling or precipitation reaction. This compound is water-insoluble and to some extent lipid-soluble. Usually the splitting and coupling reactions are performed in one step, hence the alternative name, the 'simultaneous coupling' reaction. The coupling reaction is faster than the capturing reaction of metal salt methods, so there is less chance of the primary reaction product diffusing. On the other hand, due to the lipid-solubility of azo-dyes, redistribution or recrystallization of the final coloured reaction product may occur, leading to erroneous information about the enzyme localization. When hexazotized *p*-rosanilin or hexazotized New Fuchsin are used as coupling agents, localization artefacts due to the lipid-solubility of azo-dyes are largely avoided. Another disadvantage of diazonium salt methods is the instability of diazonium salts, especially at higher pH values. Large amounts of stabilizers (sometimes up to 80 per cent) are added to commercially available diazonium salt preparations which can seriously affect histochemical reactions e.g. by enzyme inhibition.

The splitting and coupling reaction can be separated in a post-coupling method when these reactions interfere, for example when the enzyme is inhibited by diazonium salts and/or stabilizers or when the enzyme reaction has to be performed in conditions that are not favourable for the coupling reaction. A good example is the demonstration of cathepsin B activity. Cathepsin B is a cysteine proteinase which requires reducing conditions (presence of SH-groups) for its activity, but SH-groups themselves readily decompose diazonium salts. When a post-coupling method is applied, one has to be certain that the intermediate reaction product does not diffuse from the site where it was formed by the enzyme.

2.6.4 Indigogenic methods

Indigogenic methods can be used to localize non-specific esterases, cholinesterases, glycosidases, phosphatases, and proteases. The histochemical method is based on two reaction steps taking place in one and the same incubation medium. The primary reaction is once more a splitting reaction. A group, specific for the enzyme to be studied, is cleaved from an indoxyl or indolylamine derivative. Indoxyl or indolylamine is subsequently oxidized in the precipitation reaction to water-insoluble blue indigo by means of oxygen, or better, by potassium ferricyanide.

The indigogenic method can be linked with a tetrazolium salt method at neutral or alkaline pH. The indoxyl-tetrazolium salt methods are discussed in Section 2.6.6.

2.6.5 Tetrazolium salt methods

Tetrazolium salt methods for the demonstration of the activity of dehydrogenases, reductases, and oxidases are the most frequently used and the most successful methods in the enzyme histochemical repertoire. Tetrazolium salts are colourless or yellow and reasonably soluble in water; they can be reduced easily to the water-insoluble and strongly coloured formazans. Seidler (1991) extensively reviewed the underlying chemistry of tetrazolium salt reduction. The reaction chain consists of several electron transfers, the first step being the enzyme-mediated electron transfer from one substrate to the other (e.g. lactate dehydrogenase transfers electrons from lactate to NAD, thus generating pyruvate and NADH). In tetrazolium salt reactions, physiological substrates for enzymes are used. In principle all dehydrogenases, reductases, and oxidases can be demonstrated with tetrazolium salts. Usually, the electron acceptor in dehydrogenase reactions is a coenzyme such as NAD(P) or flavoprotein. The reduced coenzyme or flavoprotein reduces the tetrazolium salt to its formazan. The redox potentials of coenzyme and tetrazolium salt do not favour a fast electron flow from coenzyme to tetrazolium salt. Therefore, an exogenous electron carrier such as (1-methoxy)phenazine methosulfate or menadione is added to the incubation medium in order to speed up the electron transfer to the tetrazolium salt. When exogenous electron carriers are omitted from the incubation medium, the precipitation of formazan resembles the localization of endogenous electron transport systems such as the respiratory chain. This is of course misleading when a dehydrogenase activity should be localized, and therefore, addition of an exogenous electron carrier is essential for specific histochemical localization of the dehydrogenase to be studied.

Reductase activity can be demonstrated using reduced coenzyme (NADH or NADPH) as substrate in the absence of an exogenous electron carrier. Because of the presence of a number of reductases in most cells which can all oxidize NADH and/or NADPH more or less specifically, one should perform control incubations in the presence of specific inhibitors. For example, ferrihaemoprotein reductase (also called NADPH-cytochrome (P450) reductase) is inhibited competitively by its product NADP but not by dicumarol, whereas another reductase, NAD(P)H dehydrogenase (also called D,T-diaphorase), is inhibited by dicumarol but not by NADP. By incubating serial sections in the presence or absence of various inhibitors, the activities of different reductases in tissues can be studied.

Tetrazolium salt methods have also been used in multi-step reactions to localize enzymes that are not dehydrogenases, reductases, or oxidases, such as creatine kinase, aldolase, phosphofructokinase, and hexokinase. In these cases, the reaction product of the particular enzyme has to be a substrate for a dehydrogenase. This product is converted by the dehydrogenase with concomitant formazan production. The activity of the dehydrogenase has to be present in excess to ensure that this step is not rate-limiting, therefore the auxiliary dehydrogenase is added to the incubation medium. Of course, these multi-step reactions are rather complex, entailing the risk that an intermediate reaction product could diffuse from the original enzymic site, thus decreasing the precision of localization.

2.6.6 Indoxyl-tetrazolium salt methods

When substrates containing an indoxyl or indolylamine group are used for localization of hydrolase activity at alkaline pH, colourless dehydro-indigo can be generated instead of blue indigo. Electrons are liberated during this dehydro-indigo formation and in principle formazan will then be formed when a tetrazolium salt is added to the medium. Addition of (1-methoxy)-phenazine methosulfate, an exogenous electron carrier, also enhances electron transport in this method. The tetrazolium salt reduction proceeds faster with increasing pH, which means in practice that the method can be best applied at alkaline pH. The method has been used for the demonstration of esterases, alkaline phosphatase, and glycosidases.

2.6.7 Diaminobenzidine methods

Peroxidase and catalase activity can be demonstrated using diaminobenzidine. Both enzymes are haem proteins and oxidize their substrates using hydrogen peroxide. The histochemical method is based on the principle that the light brown water-soluble diaminobenzidine acts as hydrogen and electron donor and becomes oxidized to form a water-insoluble brown polymer. This colour can be intensified when cobalt ions or nickel ions are added to the medium. Moreover, diaminobenzidine is osmiophilic, and so the method can be used for both light and electron microscopy. The method is most widely applied in immunohistochemistry for the localization of horseradish peroxidase as a marker coupled to antibodies.

2.6.8 Fluorescence method with 5'-nitrosalicylaldehyde

Diazonium salts can be replaced by 5'-nitrosalicylaldehyde (NSA) to detect protease activity by coupling β-naphtholamine (NA) or 4-methoxy-2-naphthylamine (MNA) to NSA. These coupling products are fluorescent: green in the case of NA and yellow in the case of MNA. The coupling reaction can only take place between NSA and NA or MNA and no other naphthol derivatives, limiting this method to protease histochemistry. NSA binds non-specifically to $-NH_2$ groups in proteins, giving rise to a green fluorescent background. Although the end product is substantive (i.e. binds firmly to proteins) and does not diffuse, recrystallization of the final reaction product may reduce the precision of the localization. This recrystallization process can be avoided by using either very short incubation periods (less than 1 min) or by analysing microscopically the formation of the fluorescent product during incubation. Photomicrographs are taken during the reaction or immediately after the reaction is stopped, because preparations are not permanent. This problem occurs in many other existing fluorescence enzyme histochemical methods as well, because they yield more or less water-soluble end-products which diffuse, thus interfering with the precise localization of final reaction products.

2.6.9 Thiocholine methods

Cholinesterases hydrolyse choline esters. They can be divided into two groups: acetylcholinesterases and pseudo-cholinesterases. The former catalyse the splitting of acetylcholine into choline and water, whereas the latter catalyse the cleavage of acylcholine into choline and carboxylic acid. Acetylcholinesterase has a considerably higher affinity for acetylcholine than for choline esters. The enzyme is found in erythrocytes, perikarya of nerve cells, synapses, motor end-plates, and the conducting system in the heart.

Demonstration of acetylcholinesterase activity is performed by incubation in a medium containing acetylthiocholine as substrate, ferricyanide and Cu^{2+} ions. Thiocholine is formed upon enzyme action and this reduces ferricyanide into ferrocyanide; the latter compound reacts with Cu^{2+} ions to form water-insoluble Hatchett brown; citrate may be added as a chelator.

2.6.10 Synthesis reactions

Synthesis reactions are used to demonstrate the activity of glycosyltransferases such as glycogen phosphorylase. The reaction principle is based on the synthesis of a polyglycan such as glycogen that then can be visualized. Glycogen phosphorylase is demonstrated by adding a high concentration of glucose-1-phosphate to the incubation medium. This forces the enzyme to synthesize glycogen instead of breaking it down. Endogenous glycogen is used as a primer. The synthesized polyglycan is then stained with the periodic acid–Schiff technique. Control sections are stained for endogenous polyglycan. The difference in staining between the test and controls represents the activity of the glycosyltransferase.

2.6.11 Natural chromophores

The physiological product generated by some enzymes can be used directly for histochemical purposes because it is coloured and water-insoluble. The best-known example is that of L-DOPA oxidase, which is involved in the formation of various melanin pigments. These are complex, high molecular weight polymers formed from 3,4-dihydroxyphenylalanine (L-DOPA).

The histochemical reaction for the demonstration of cytochrome P450 also fits into this category. The reduced form of cytochrome P450 produces a characteristic absorbance band at 450 nm when carbon monoxide is bound to the enzyme molecule. Therefore, serial sections are incubated either in air or in a medium saturated with carbon monoxide either in the presence or absence of sodium dithionite. The latter compound reduces cytochrome P450 and therefore, in the presence of carbon monoxide, the absorbance at 450 nm can be monitored. All other serial sections serve as controls.

2.6.12 Concluding remarks

The reliability and precision of localization of enzyme activity by a final water-insoluble coloured reaction product depend on a number of methodological aspects. Diffusion has already been mentioned several times. It always occurs to a certain extent because capture reactions are never completely efficient, and therefore intermediate reaction products can diffuse before being captured. Several methods have been developed to minimize diffusion. For example, sections or cells are incubated in a viscous medium, or a semipermeable membrane is interposed between section and (gelled) incubation medium. The most frequently used method to prevent diffusion of enzyme molecules from unfixed sections is addition of polyvinyl alcohol to the reaction medium. Polyvinyl alcohol is an inert macromolecule and at the proper concentration small molecules such as substrates and colour reagents can diffuse freely from the medium into the specimen, whereas diffusion of macromolecules is hampered by the meshwork of polyvinyl alcohol molecules in the medium (Van Noorden and Vogels 1989). Morphology of unfixed tissue preparations is preserved considerably better after incubation in a polyvinyl alcohol-containing medium than in unfixed tissue incubated in aqueous medium without polyvinyl alcohol. This is also true for the morphology of cryostat sections at the ultrastructrual level after incubation with the semipermeable membrane technique (Fig. 2.4). In the presence of polyvinyl alcohol, high concentrations of capturing reagents can be built up at sites of enzyme activity. These high concentrations help to prevent diffusion of intermediate reaction products from the enzyme site as the capturing reaction is performed faster.

For proper enzyme histochemistry, unfixed cryostat sections or unfixed cell preparations should be used. Chemical fixation of tissue rapidly inactivates enzymes. Only a few enzymes, such as certain phosphatases and oxidases, can withstand fixatives, especially the fixatives that are used routinely in classical histology, e.g. 4 per cent glutaraldehyde. Very low concentrations such as 0.025 per cent glutaraldehyde have been applied successfully to enzyme histochemistry, but other mild fixatives such as carbodiimide should be tested as well (Van Duijn 1991).

26 *Enzyme histochemistry: a laboratory manual of current methods*

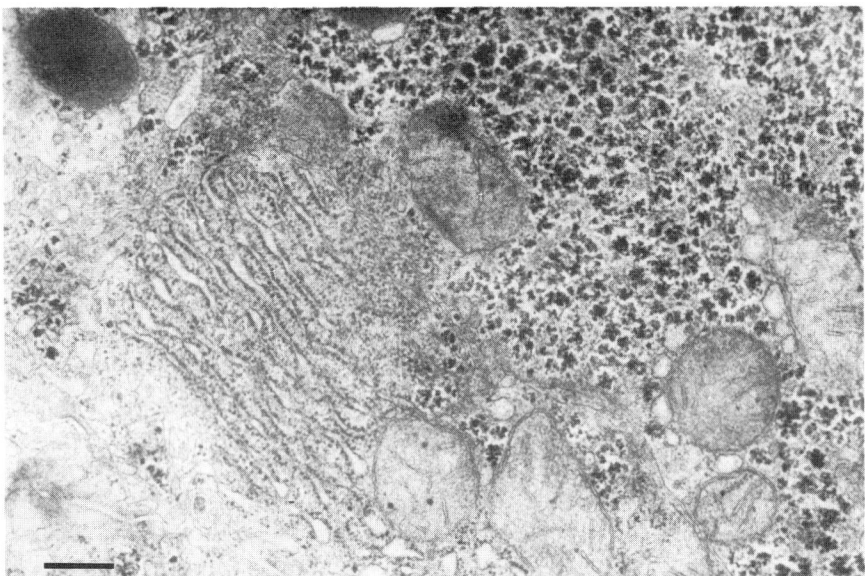

Fig. 2.4. Electron micrograph of part of a hepatocyte in a cryostat section of unfixed liver tissue. The cryostat section was attached to a semipermeable membrane, subsequently fixed and embedded. Ultrathin sections were cut according to routine electron microscopical procedures. Note the excellent morphology that can be obtained with cryostat sections of unfixed tissue. Bar, 50 nm.

Another possibility which offers some protection to enzymes against inactivation during fixation is preincubation of the tissue with its substrate. The substrate covers the active site and protects it against chemical damage. Substrate protection has been successfully used in a number of enzyme reactions and deserves further investigation (Van Duijn 1991; Van Noorden and Butcher 1991).

2.7 Unit of comparison of biochemical and quantitative histochemical data

In 1961, Glick predicted that, particularly in fundamental metabolic studies, efforts in the next decades would be directed towards closing the gap between biochemistry and histochemistry by trying to localize what can be quantified and to quantify what can be localized. This gap has now been closed to a large extent. Localization of what can be quantified is routinely performed with microchemical assays (Lowry and Passonneau 1972; Outlaw *et al.* 1985; for a list of enzymes assayed with microchemical methods, see Van Noorden and

Butcher 1991) and quantification of what can be localized is performed with cytophotometry.

Microchemical assays are scaled-down biochemical assays in which enzyme reactions are analysed in one individual cell or a small group of cells dissected from a freeze-dried cryostat section. A serial section is used for orientation and to provide morphological information (Galjaard 1980; Katz 1989). These methods are very accurate and sensitive but they are rather time-consuming and require sophisticated equipment and are not described here. Cytophotometric analysis of enzyme reactions in tissue sections is relatively simple; the equipment needed is a cryostat (preferably motor-driven to ensure constant cutting speed and thus constant section thickness) and either a scanning and integrating cytophotometer or an image analysing system (Van Noorden and Butcher 1991). For the advantages of image analysis over cytophotometry, see Section 2.5.

Now that the gap has been closed, it is particularly useful to focus on the comparison of activity and kinetic parameters of (partially) purified enzyme molecules in diluted solutions as analysed in biochemical assays with those of enzymes in their cellular environment as analysed histochemically. This direct comparison can be performed for those histochemical methods that are based on the formation of final reaction product with known molar extinction coefficients in their precipitated forms. They are listed in Table 2.3. In Method 3, the calculation procedures are given for this direct comparison on the basis of wet weight of tissue. For both the biochemical and histochemical assays the data are expressed as moles of substrate converted per unit time per unit wet weight of tissue.

Other references for enzyme activity, such as DNA or protein content or unit dry weight, may be more useful. Usually, reference to DNA content is the best reference measure for enzymes in cells. This is shown in Table 2.4 (Van Noorden and Butcher 1991).

Irrespective of the reference measure taken, tetrazolium salt methods can

Table 2.3. *Molecular extinction coefficients of final reaction products precipitated in tissue sections or cells*

Final reaction product	Wavelength (nm)	ε
Formazan of neotetrazolium	585	7 000[a]
Formazan of nitro BT	585	16 000[a]
Formazan of tetranitro BT	557	19 000[a]
Lead sulfide	450	5 700[b]
Poly(diaminobenzidine)	480	5 500[b]

[a] With respect to moles substrate converted in redox reactions.
[b] With respect to moles phosphate liberated in phosphatase reactions.

now be used for direct comparison of histochemically and biochemically obtained data on enzyme reactions.

Table 2.4. *Cytophotometric analysis of glucose-6-phosphate dehydrogenase activity per unit wet weight of tissue per unit time (μmoles glucose-6-phosphate converted per g wet weight per min) and per cell (fmoles glucose-6-phosphate converted per cell per min) in three different tissues (after Van Noorden and Butcher 1991). The tissues consist of cells with different cytoplasmic volumes*

Tissue	Cells per unit volume $\times 10^8 \cdot cm^{-3}$	Activity per unit wet weight per unit time	Activity per cell per unit time
Tissue I	8.8	55.6	62.9
Tissue II	7.3	46.5	64.0
Tissue III	3.4	23.8	70.3

Method 3. *Conversion of cytophotometric arbitrary machine units into mean integrated absorbance and absolute units of enzyme activity (Van Noorden and Gossrau 1991)*

1. Use a series of neutral grey filters of known absorbance to make calibration curves of the arbitrary cytophotometric machine units and absorbance.
2. Apply this curve for conversion into mean integrated absorbance. The conversion factor depends on the area measured (or in other words, the number of measuring points of a scanning and integrating cytophotometer or the number of pixels of an image). For example, the conversion factor for calculating mean integrated absorbance from arbitrary machine units of a Vickers M85 cytophotometer is 0.0164 for the A2 mask and 0.0072 for the A3 mask, independent of the objective.
3. Use the molar extinction coefficients shown in Table 2.3 to convert mean integrated absorbance into concentration of final reaction product in the Lambert–Beer equation, $A = \varepsilon c d$, in which A is the mean integrated absorbance (dimensionless), ε is the molar extinction co-efficient (litres·mole^{-1}·cm^{-1}), c is the concentration (moles/litre), and d is the section thickness (in cm). See Table 2.5 for an example.
4. Calculate how many moles of substrate have been converted for the generation of 1 mole final reaction product. For example, 1 mole tetranitro BT-formazan is generated when 2 moles substrate are converted, because two electrons are liberated per substrate molecule, while the diformazan takes up four electrons when fully reduced. However, the values for formazans in Table 2.3 are given per two electrons taken up at the isobestic wavelength (Van Noorden and Butcher 1991).
5. Calculate the amount of substrate converted per unit volume of wet tissue (say, per cm^3) and per unit time (say, per min) (See Table 2.5).

Table 2.5. *Example of the conversion procedure of arbitrary cytophotometric machine units to absolute units of enzyme activity as presented in Method 3 with the example of the conversion of glucose-6-phosphate (G6P) by glucose-6-phosphatase activity (after Jonges et al. 1992)*

	Value	Conversion factor
1. Arbitrary machine units: (Vickers M85A, A2 mask, 2 s scan)	50	
		$\times 0.0164$
2. Mean integrated absorbance (MIA):	0.82	
		$\div 5500^a \times \dfrac{10\,000^b}{8}$
3. mmoles G6P·ml^{-1}·5 min^{-1}:	0.186	
		$\div 5^c$
4. mmoles G6P·ml^{-1}·min^{-1}:	0.037	
		$\times 1000$
5. µmoles G6P converted·g wet weight of tissue^{-1}·min^{-1}:	37	

[a] Molar extinction coefficient of poly(diaminobenzidine) with respect to phosphate liberated (see Table 2.3).
[b] Section thickness was 8 µm or 8/10 000 cm.
[c] Incubation lasted 5 min.

2.8 Metabolic fluxes

The latest innovation in the histochemical analysis of cellular metabolism has been introduced by Mueller-Klieser and co-workers (1988). The authors described a new method for determining substrate concentrations in cryostat sections of tissues using an image analysis system. The cryostat section is heat-treated to destroy all enzyme activity. The substrate to be measured is converted by an auxiliary enzyme that is present in a second cryostat section made from a gelatin solution of the enzyme together with the compounds necessary to convert all substrate present in the tissue section. The second gelatin cryostat section is sandwiched with the first one and the reaction is started. The enzyme reaction is coupled to a bioluminescence-generating system containing luciferase and the photons produced are measured by a single-photon-imaging system. So far, concentrations of lactate, ATP, and glucose have been determined in this way. Conversion into absolute concentrations of substrate is possible because model cryostat sections with known concentrations of substrate can be used for calibration. The spatial resolution is 10–100 µm, depending on microscopical magnification, intensity of light emission, and section thickness, whereas substrate concentrations as low as 0.5 mM can be measured. These *in situ* substrate measurements in combination with the determination of *in situ* kinetic parameters of enzymes enable the calculation of local metabolic fluxes.

Metabolic fluxes are direct parameters of cellular metabolic activity (Newsholme et al. 1980).

In Table 2.6, an example is given of the calculation of metabolic fluxes when the V_{max}, K_M, and the physiological substrate concentration are known. It shows that the conversion of glucose-6-phosphate into glucose by glucose-6-phosphatase in female rat liver lobules after feeding and 24 hr starvation is rather constant in liver tissue despite the fact that V_{max} and K_M values differ significantly in the different areas of the liver lobules and after feeding or fasting.

Table 2.6. *Kinetic parameters K_M and V_{max}, local substrate concentrations and flux rates for the conversion of glucose-6-phosphate (G6P) to glucose and P_i by glucose-6-phosphatase in pericentral and periportal zones of female rat liver lobuli either in the fed or fasted state (after Jonges et al. 1992)*

Parameter	Pericentral zone Fed	24 h Fasted	Periportal zone Fed	24 h Fasted
K_M (mM)	0.41	0.98	1.24	1.87
V_{max} (µmoles G6P converted·min^{-1}·g liver^{-1}	6.36	15.00	15.00	41.82
[Glucose-6-phosphate] (µmoles G6P·g liver^{-1})	0.104	0.078	0.093	0.075
G6P flux (µmoles G6P·min^{-1}·g liver^{-1})	1.29	1.11	1.05	1.6

2.9 Immunocytochemical detection of enzyme molecules and detection of mRNAs coding for enzymes by *in situ* hydridization versus *in situ* localization of enzyme reactions

Immunocytochemistry has developed very rapidly over the past decades and it is an extremely important approach in cell biological research and diagnostic pathology (for a review of methods, see Polak and Van Noorden 1991). A large number of enzymes can presently be detected by immunohistochemical means (see e.g. Lamers et al. 1989). Similarly, mRNAs can be detected with *in situ* hybridization techniques, including mRNAs coding for enzymes proteins (see e.g. Moorman et al. 1990).

Some enzymes are synthesized in an inactive pro- or pre-pro-form which has to be processed at some stage in order to become active, whereas other enzymes such as proteases may be present in an inactive form complexed with their natural inhibitors. Therefore it is of great interest to study regulation mechanisms in cells at different levels such as at the transcriptional level where mRNA is formed, the translational level when ((pre-)pro-)enzymes are synthesized as encoded by the mRNA, or at the post-translational level when the enzyme molecules are modified (Fig. 2.5). Modifications are performed e.g. by proteolytic cleavage of the pro-form to generate the active form, phosphorylation, disruption of the binding of the enzyme and its inhibitor, or disruption of the binding of the enzyme to structural elements (e.g. the binding of glucose-6-phosphate dehydrogenase to the cytoskeleton (Swezey and Epel 1986)). These post-translational regulatory processes can be very fast (a matter of minutes or seconds) and can be due to phenomena such as changes in cellular Ca^{2+} fluxes, oxygen radical formation, or proteolytic activity.

The methodology for the researcher now contains probes for *in situ* hybridization to detect specific mRNA sequences encoding the enzyme, monoclonal or polyclonal antibodies for the specific immunocytochemical detection of the total amount of active and inactive forms of an enzyme, and substrates and other reagents for the detection of the activity of enzymes (Fig. 2.5). Fig. 2.6 shows an example of such an analysis for glutamate

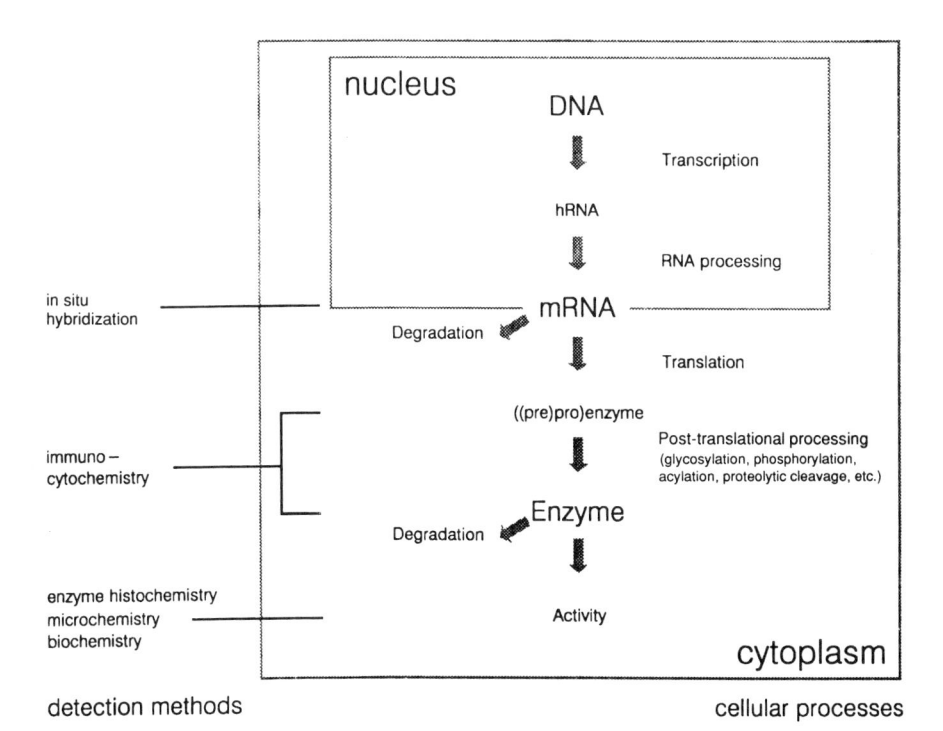

Fig. 2.5. Scheme of the different cellular processes between transcription of a gene in the nucleus and the activity of an enzyme encoded by that gene together with the different (histochemical) detection methods that can be used to analyse the regulation mechanisms of these cellular processes. See Fig. 2.6 for an example.

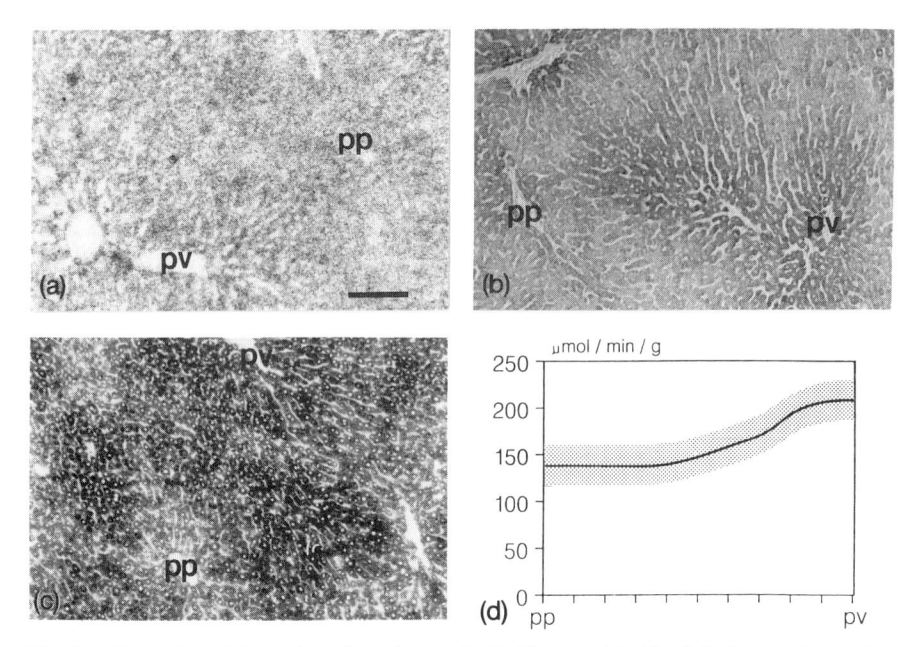

Fig. 2.6. Examples of the various detection methods, illustrated in Fig. 2.5, that can be used to analyse cellular processes involved in regulation of enzyme activity. Glutamate dehydrogenase (NAD$^+$ dependent; EC 1.4.1.2) in rat liver is used as the example. (a) Demonstration of mRNA encoding glutamate dehydrogenase. (b) Demonstration of molecules of glutamate dehydrogenase using immunocytochemistry. (c) Demonstration of activity of glutamate dehydrogenase using enzyme histochemistry. (d) Quantification of glutamate dehydrogenase activity using microchemistry. The solid line represents the distribution pattern of activity in the liver lobulus from the portal tract (pp) to the central vein (pv); the shaded area represents the standard deviation. Note that the distribution patterns of the various molecules in the liver lobulus show highest amounts in pericentral (pv) zones and lower amounts in periportal (pp) zones. However, the width of the zones do not correlate entirely (particularly the amount of protein molecules (b) in comparison with the amount of mRNA (a) and activity (c and d)). Panel (a) was kindly provided by Dr A. F. M. Moorman and Panel (b) by Prof. Dr W. H. Lamers of the Laboratory of Anatomy and Embryology of the University of Amsterdam. Panel (d) was redrawn from Maly and Sasse (1991). Bar, 100 μm.

dehydrogenase. It is shown that the patterns for mRNA, the enzyme protein, and the activity of the enzyme do not entirely correlate. Although the reasons for this discrepancy are not yet known, these combined histochemical approaches will undoubtedly become increasingly important for the analysis of cellular metabolism.

References

Angermueller, S. (1989). Peroxisomal oxidases: cytochemical localization and biological relevance. *Progress in Histochemistry and Cytochemistry*, **20**(1), 1–65.

Dixon, M. and Webb, E. C. (1979). *Enzymes*, 3rd edn. Longman, London.
Fahimi, H. D. (1980). Qualitative cytological criteria for the validation of enzyme histochemical techniques. In *Trends in Enzyme Histochemistry and Cytochemistry* (ed. D. Evered and M. O'Connor), pp. 33–51. Excerpta Medica, Amsterdam.
Galjaard, H. (1980). Quantitative cytochemical analysis of (single) cultured cells. In *Trends in Enzyme Histochemistry and Cytochemistry* (ed. D. Evered and M. O'Connor), pp. 161–80. Excerpta Medica, Amsterdam.
Glick, D. (1961). *Quantitative Chemical Techniques of Histo- and Cytochemistry*, Vol. 1 and 2. Wiley-Interscience, New York.
Groen, A. K., Van Der Meer, R., Westerhoff, H. V., Wanders, R. J. A., Akerboom, T. P. M., and Tager, J. M. (1982). Control of metabolic fluxes. In *Metabolic Compartmentation* (ed. H. Sies), pp. 9–37. Academic Press, New York.
Hulstaert, C. E., Kalicharan, D., and Hardonk, M. J. (1989). Cytochemical demonstration of phosphatases with the cerium method. In *Techniques in Diagnostic Pathology*, Vol. 1 (ed. G. R. Bullock), pp. 133–50. Academic Press, New York.
James, J. (1983). Developments in photometric techniques in static and flow systems from 1960 to 1980: a review, including some personal observations. *Histochemical Journal*, **15**, 95–110.
James, J. and Tanke, H. J. (1991). *Biomedical Light Microscopy*. Kluwer Academic Publishers, Dordrecht.
Jonges, G. N., Van Noorden, C. J. F., and Gossrau, R. (1990). Quantitative histochemical analysis of glucose-6-phosphatase activity in rat liver using an optimized cerium-diaminobenzidine method. *Journal of Histochemistry and Cytochemistry*, **38**, 1413–19.
Jonges, G. N., Van Noorden, C. J. F., and Lamers, W. H. (1992). In situ kinetic parameters of glucose-6-phosphatase in the rat liver lobulus. *Journal of Biological Chemistry*, **267**, 4878–81.
Katz, N. R. (1989). Methods for the study of liver cell heterogeneity. *Histochemical Journal*, **21**, 517–29.
Lamers, W. H., Moorman, A. F. M., and Charles, R. (1989). The metabolic lobulus, a key to the architecture of the liver. In *Hepatocyte Heterogeneity and Liver Function* (ed. J. J. Gumucio). *Cell Biological Reviews*, **19**, 5–25.
Lawrence, G. M., Beesley, A. C. H., Mason, G. I., Thompson, M., Walker, D. G., and Matthews, J. B. (1989). A comparison of histochemically and biochemically determined kinetic parameters for brain hexokinase Type I. In *Proceedings of EMAG-MICRO 89*, Vol. 2 (ed. H. Y. Elder and P. J. Goodhew), pp. 667–70. Institute of Physics, Bristol.
Lawrence, G. M., Beesley, A. C. H., Mason, G. I., Deacon, E. M., and Matthews, J. B. (1990). TV image analysis and the quantification of immunocytochemical and enzyme histochemical staining. In *Transactions of the Royal Microscopical Society*, Vol. 1 (ed. H. Y. Elder), pp. 519–24. Adam Hilger, Bristol.
Lowry, O. H. and Passonneau, J. V. (1972). *A Flexible System of Enzymatic Analysis*. Academic Press, New York.
Maly, I. P. and Sasse, D. (1991). Microquantitative analysis of the intra-acinar profiles of glutamate dehydrogenase in rat liver. *Journal of Histochemistry and Cytochemistry*, **39**, 1121–4.
Moorman, A. F. M., De Boer, P. A. J., Das, A. T., Labruyere, W. T., Charles, R.,

and Lamers, W. H. (1990). Expression patterns of mRNAs for ammonia-metabolizing enzymes in the developing rat: the ontogenesis of hepatocyte heterogeneity. *Histochemical Journal,* **22,** 457–68.

Mueller-Klieser, W., Walenta, S., Paschen, W., Kallinowski, F., and Vaupel, P. (1988). Metabolic imaging in microregions of tumors and normal tissues with bioluminescence and photon counting. *Journal of the National Cancer Institute,* **80,** 842–8.

Newsholme, E. A., Crabtree, B., and Zammit, V. A. (1980). Use of enzyme activities as indices of maximum rates of fuel utilization. In *Trends in Enzyme Histochemistry and Cytochemistry* (ed. D. Evered and M. O'Connor), pp. 245–58. Excerpta Medica, Amsterdam.

Outlaw, W. H., Springer, S. A., and Tarczynski, M. C. (1985). Histochemical technique. A general method for quantitative enzyme assays of single cell 'extracts' with a time resolution of seconds and a reading precision of femtomoles. *Plant Physiology,* **77,** 659–66.

Pette, D. and Wimmer, M. (1980). Microphotometric determination of enzyme activities in cryostat sections by the gel film technique. In *Trends in Enzyme Histochemistry and Cytochemistry* (ed. D. Evered and M. O'Connor), pp. 121–34. Excerpta Medica, Amsterdam.

Polak, J. and Van Noorden, S. (1991). *An Introduction to Immunocytochemistry: Current Techniques and Problems.* Oxford University Press, Oxford.

Raap, A. K. (1986). Localization properties of fluorescence cytochemical enzyme procedures. *Histochemistry,* **84,** 317–21.

Seidler, E. (1991). The tetrazolium-formazan system: design and histochemistry. *Progress in Histochemistry and Cytochemistry,* **24**(1), 1–86.

Stoward, P. J. (1980). Criteria for the validation of quantitative enzyme histochemical techniques. In *Trends in Enzyme Histochemistry and Cytochemistry* (ed. D. Evered and M. O'Connor), pp. 11–31. Excerpta Medica, Amsterdam.

Swezey, R. R. and Epel, D. (1986). Regulation of glucose-6-phosphate dehydrogenase activity in sea urchin eggs by reversible association with cell structural elements. *Journal of Cell Biology,* **103,** 1509–15.

Van Duijn, P. (1991). Model systems. Principles and practice of the use of matrix-immobilized enzymes for the study of the fundamental aspects of cytochemical enzyme methods. In *Histochemistry. Theoretical and Applied,* Vol. 3, 4th edn. (ed. P. J. Stoward and A. G. E. Pearse), pp. 433–472. Churchill Livingstone, Edinburgh.

Van Noorden, C. J. F. (1989). Principles of cytophotometry in enzyme histochemistry and validity of the reactions. *Acta Histochemica,* **Supplement 37,** 21–35.

Van Noorden, C. J. F. (1991). Assessment of lysosomal function by quantitative histochemical and cytochemical methods. *Histochemical Journal,* **23,** 429–35.

Van Noorden, C. J. F. and Butcher, R. G. (1991). Quantitative enzyme histochemistry. In *Histochemistry. Theoretical and Applied,* Vol. 3, 4th edn. (ed. P. J. Stoward and A. G. E. Pearse), pp. 355–432. Churchill Livingstone, Edinburgh.

Van Noorden, C. J. F. and Gossrau, R. (1991). Quantitative histochemical and cytochemical assays. In *Histochemical and Immunohistochemical Techniques. Applications to Pharmacology and Toxicology* (ed. P. H. Bach and J. R. J. Baker), pp. 119–45. Chapman and Hall, London.

Van Noorden, C. J. F. and Vogels, I. M. C. (1989). Polyvinyl alcohol and other tissue protectants in enzyme histochemistry. A consumer's guide. *Histochemical Journal,* **21**, 373–9.

Van Noorden, C. J. F., Dolbeare, F., and Aten, J. (1989). Flow cytofluorometric analysis of enzyme reactions based on quenching of fluorescence by the final reaction product: detection of glucose-6-phosphate dehydrogenase deficiency in human erythrocytes. *Journal of Histochemistry and Cytochemistry,* **37**, 1313–18.

Wilkinson, G. N. (1961). Statistical estimations in enzyme kinetics. *Biochemical Journal,* **80**, 324–32.

Zollner, H. (1989). *Handbook of Enzyme Inhibitors.* V.C.H. Gesellschaft, Weinheim.

3 Preparation techniques

3.1 Tissue and cell preparations

There are basically four options for pretreatment of tissues and cells for light microscopical enzyme histochemical studies:

1. Snap freezing of cell preparations or small blocks of tissue as soon as they are obtained and storage at −80 °C or in liquid nitrogen at −196 °C (see Section 3.1.1).
2. Chemical fixation of cells and tissues as soon as they are obtained and subsequent freezing (see Section 3.1.2).
3. Chemical fixation and infiltration with glycol methacrylate resin monomer at 4 °C and polymerization at 4 °C (see Section 3.1.3).
4. Quenching of cells and tissues in liquid nitrogen as soon as they are obtained, followed by freeze-substitution at −35 °C, infiltration with glycol methacrylate resin monomer at 4 °C, and polymerization at 4 °C (Section 3.1.4).

3.1.1 Snap freezing

It is a general misconception that cryostat sections or cell preparations kept at −80 °C show bad morphology due to ice crystal formation (see e.g. Lojda *et al.* 1979; Pretlow *et al.* 1987; Murray and Ewen 1989). When tissue blocks or cell preparations are carefully frozen according to Method 4 and Section 3.3, stored at low temperature (less than −70 °C) and cryostat sectioning is performed at less than −20 °C according to Section 3.2.1.1, even the ultrastructural morphology can be excellent irrespective of the type of tissue or cell involved (Fig. 2.4). However, these cryostat sections or cell preparations have to be incubated very carefully for the demonstration of enzyme activity, according to the Methods in Chapter 4, because morphology almost always deteriorates during incubation. When these aspects are taken into consideration, the use of unfixed tissue sections or cell preparations is the best option for enzyme histochemical purposes. The advantages are that enzyme activity is not lost due to fixation or embedding; a wide range of tissues and cells from many animals and plants can be chilled by the same technique; and high quality morphology of tissues and cells is obtained when incubations are performed according to standard procedures (Fig. 2.4).

Method 4. *Snap freezing of tissues*

1. As soon as the tissue is obtained, cut it into blocks up to 5 mm thick.
2. Put the pieces of tissue into small screw capped plastic incubation vials and close the vials.
3. Slowly immerse the vial containing the piece of tissue in liquid nitrogen and keep it there for at least 5 min.
4. Store the vial with tissue in a biological freezer at −80 °C or in liquid nitrogen at −196 °C.

3.1.2 Fixation and snap freezing

Fixation and/or mounting of tissues and cells lead to lesser or greater degree of loss of enzyme activity and thus to limited information. When fixation is thought to be necessary it is essential to select fixation conditions that guarantee limited loss of enzyme activity and best morphology. Such fixation procedures (Van Noorden and Hulstaert 1991) are given in Methods 5–7.

Method 5. *Perfusion fixation of whole animals*

1. Anaesthetize animal with Nembutal.
2. Administer 0.45 mg sodium nitrite and 250 IU heparin in 250 µl of a 0.9 per cent sodium chloride solution intravenously into tail vein to prevent contraction of blood vessels.
3. Open abdomen and cut the chest open at both sides of the sternum.
4. Remove pericardium from heart.
5. Insert needle of perfusion system into left ventricle.
6. Perfuse with 2.5 per cent glutaraldehyde in 100 mM cacodylate buffer (pH 7.4; temperature of fixative should be 20 °C). Flow rate of perfusion medium: 2.5 ml/min*.
7. Open right atrium to allow body fluids and fixative to leave the body after perfusion.
8. Perfuse for 1–10 min depending on stiffening of the body.
9. Perfuse for 5 min with 100 mM cacodylate buffer (pH 7.4).
10. Remove organs; cut and freeze tissue as in Method 4.

* Flow rate is given for rats and mice. For larger or smaller animals this value needs adjustment.

Method 6. *Perfusion fixation of organs*

1. Perform perfusion with a pump with a flow rate of 13–15 ml/min at room temperature.
2. Perfuse for 1 min with rinsing medium consisting of 2 per cent polyvinyl pyrrolidone (PVP, mol. wt 40 000), 75 mM sodium nitrite in 100 mM cacodylate buffer (pH 7.4). PVP compensates for the colloid-osmotic pressure; sodium nitrite prevents contraction of blood vessels.
3. Perfuse with 2 per cent glutaraldehyde in 100 mM cacodylate buffer (pH 7.4) for 5 min.
4. Perfuse for 5 min with 6 per cent sucrose in 100 mM cacodylate buffer (pH 7.4).
5. Cut and freeze tissue as in Method 4.

Method 7. *Immersion fixation*

1. Cut tissue into blocks that are as small as possible.
2. Immerse tissue in 1–2 per cent glutaraldehyde and 6 per cent sucrose in 100 mM cacodylate buffer (pH 7.2) for 45 min at 4 °C.
3. Rinse tissue blocks three times in 100 mM cacodylate buffer (pH 7.2).
4. Immerse blocks in a 7.5 per cent agar solution and let it solidify.
5. Cut and freeze tissue as in Method 4.

3.1.3 Fixation and embedding

Embedding usually results in the loss of most enzyme activity. Nowadays, enzyme incubations are performed before embedding for most ultrastructural purposes (Hulstaert *et al.* 1989; Van Noorden and Hulstaert 1991). The only reliable method of embedding of fixed tissue that yields sections in which at least some enzyme activity can be detected using light microscopy has been developed by Pretlow *et al.* (1987) and Murray and co-workers (Murray *et al.* 1988; Murray and Ewen 1989, 1990) (see Method 8). The method is based on embedding in glycol methacrylate at 4 °C to dissipate the heat generated during polymerization which would otherwise lead to rapid enzyme inactivation. However, to give an idea of loss of enzyme activity after embedding, the activity of most enzymes can be demonstrated only after incubations of from 30 min to 4 h at 37 °C, whereas activity of the same enzymes in unfixed cryostat sections can be visualized after incubations of 5–15 min at 37 °C (see Methods in Chapter 4).

Method 8. *Embedding fixed tissue in glycol methacrylate*

1. Dehydrate fixed tissue blocks in 100 per cent acetone at −20 °C for 6 h.
2. Transfer blocks to a 1:1 (v/v) mixture of acetone and resin monomer solution (a 60:40 (v/v) mixture of glycol methacrylate and butanediol monoacrylate) and incubate for 1 h at 4 °C with continuous stirring.
3. Transfer blocks to the resin monomer solution and incubate for 4 h at 4 °C.
4. Impregnate blocks with the resin monomer solution containing 0.6 per cent benzoyl peroxide for 15 min at 4 °C.
5. Soak tissue blocks for 10 min at 4 °C with continuous stirring in a 600:1 (v/v) mixture of resin monomer solution and N,N-dimethylaniline containing 0.6 per cent benzoyl peroxide.
6. Transfer blocks to moulds and add the same mixture.
7. Allow resin to polymerize for 8 h at 4 °C.
8. Store blocks in air-tight containers at 4 °C or −20 °C.

3.1.4 Freeze-substitution and embedding

Best results with glycol methacrylate-mounted tissues are obtained when tissues are freeze-substituted without fixation as described in Method 9.

Method 9. *Embedding freeze-substituted tissue in glycol methacrylate*

1. As soon as tissue is obtained, cut it into blocks up to 5 mm thick and snap freeze as described in Method 4.
2. Freeze-substitute unfixed blocks of tissue for 18 h at −35 °C by immersing the frozen tissue blocks in acetone.
3. Embed the unfixed tissue blocks in glycol methacrylate according to steps 2–8 of Method 8.

3.2 Preparation of sections

Tissue sectioning is another critical step in enzyme histochemical procedures, especially when unfixed frozen material is used. When cryostat sections are prepared from fixed material, measures have to be taken to ensure that the sections adhere to the slides during subsequent steps in the procedure.

3.2.1 Cryostat sections

Cryostat sections are prepared from frozen tissue that is usually kept at −80 °C after being snap frozen (see Section 3.1.1). In order to minimize tissue damage it is essential that the tissue is only warmed up once from −80 °C to −30 °C between storage and picking up of the sections on glass slides in the cryostat. Thawing and freezing processes are detrimental for the morphology due to formation of ice crystals and, therefore mounting of frozen tissue on metal chucks that are placed in the cryostat cabinet has to be carried out carefully according to Method 10.

3.2.1.1 Unfixed cryostat sections

Cut sections and pick up sections on glass slides according to Method 10.

Method 10. *Cutting unfixed cryostat sections*

1. Attach tissue block to metal chuck in the cryostat cabinet using TissueTek at an ambient temperature between −20 °C and −30 °C, without allowing the tissue block to be warmed up by the liquid TissueTek.
2. Start sectioning when the block is trimmed to the desired level in the tissue block.
3. Cut sections with the desired thickness at a slow but constant speed.
4. When a section is cut correctly it remains flat on the microtome knife under the anti-roll plate.
5. When all causes of bad sectioning discussed below are taken into consideration, it is possible to cut excellent sections with a thickness of 8–10 µm; sections can be cut at any required thickness between 2 and 30 µm.

General causes of bad sectioning and poor morphology and their solutions

- The tissue block is not adjusted to the cryostat cabinet temperature. Too low a tissue temperature results in bad sectioning: the tissue and the cabinet temperature should be similar, both being between −20 °C and −30 °C. When lipid-rich tissue has to be cut, the optimum temperature is −35 °C.
- The knife and/or knifeholder are not firmly fixed. All appropriate screws must be checked and tightened.
- The anti-roll plate is not properly adjusted. The anti-roll plate of transparent Perspex prevents sections from curling up as they are cut. Small nylon screws at the side allow the adjustment of the gap between the anti-

roll plate and the knife. The section should slide between the knife and the anti-roll plate. The top edges of the anti-roll plate and the knife facet should be aligned carefully: the edge of the anti-roll plate should be marginally higher than the edge of the knife facet. When it is lower than the cutting edge, the section will curl up and not glide between knife and plate; when the plate edge is too high, it will touch the block while cutting.
- The angle of the knife to the tissue surface should generally be 15°, but this depends on the tissue to be cut and an angle as low as 10° could be the best option (see Section 5.4).

Other problems with sectioning and their solutions

Sections are not the same size as the tissue surface
- The speed of sectioning should be constant and not too high. Motor-driven cryostats are best because a constant low cutting speed ensures constant section thickness which is essential for quantitative purposes. If sections are too small, the speed is too high; if sections are too large the speed is too low.

Sections are not obtained as a whole but break into two or more pieces
- The knife temperature is too high. Ideally the knife should be cooler than the cabinet and tissue block in order to dissipate the heat generated by cutting (Chayen *et al*. 1973). This can be achieved by cooling the knife by packing dry ice around it.
- The knife edge or anti-roll plate edge contain one or more burrs. The knife should be sharpened or the plate should be burnished. It is also possible that small hardened pieces of tissue are attached to the edge of the knife or plate. Cleaning with acetone-wetted paper tissue solves this problem.
- The tissue block should be positioned in the cryostat in such a way that the smallest edge of the block faces the edge of the microtome.
- Tissue blocks containing epithelium at an edge should be positioned with the epithelium on the side and not at the top or bottom facing the knife, otherwise the epithelium is either damaged or lost from the section.

The quality of sections increasingly deteriorates over hours, days or weeks
- This is often caused by the cabinet temperature being too high, despite an intact cooling unit, or by a rapidly rising temperature during sectioning. In this case the cryostat must be defrosted and cleaned because the refrigerating elements are covered with ice. At the same time the microtome should be cleaned and oil residues removed. Before the microtome is installed again in the cryostat cabinet it should be entirely dry (drying at least overnight, preferably at an elevated temperature). Then the

appropriate points should be greased with special cryostat oil. When the microtome is installed and the cooling unit switched on, it takes another night before the cabinet is adjusted to the desired temperature.

- In order to achieve good sectioning, it is necessary that the sectioning surface of the tissue block is smooth. This can be done by levelling the top of the block with a razor blade or (better) by rapidly cutting 30 μm thick sections without using the anti-roll plate until the desired level in the block is reached. This can be checked by picking up a section on a glass slide, staining it for 1 min with an aqueous solution of 0.5–1.0 per cent (w/v) Methylene Blue or a Giemsa solution, rinsing it with water and examining it under the microscope.

3.2.1.2 Fixed cryostat sections

Cryostat sections obtained from frozen blocks of fixed tissue (see Section 3.1.2) can be cut in a similar way to that described for unfixed sections in Section 3.2.1.1. However, it is far easier to obtain sections from fixed than from unfixed material and therefore, all the points mentioned for sectioning of unfixed sections do not necessarily have to be carried out under the same rigid conditions. However, (parts of) fixed sections are easily lost from glass slides during incubation; this can largely be prevented by using pretreated glass slides (see Method 11).

Method 11. *Coating of glass slides with poly-L-lysine or gelatin*

1. Prepare a solution of either 0.1 per cent (w/v) poly-L-lysine in distilled water or 1 per cent (w/v) gelatin in distilled water containing 0.1 per cent (w/v) potassium chromium sulfate.
2. Put glass slides in either of the solutions. For gelatin-coating, keep the solution at 70 °C.
3. Let the glass slides dry standing up, at room temperature for poly-L-lysine or at 70 °C for gelatin.
4. Store slides at −20 °C.

3.2.2 Glycol methacrylate sections

Sections of tissues mounted in glycol methacrylate can be cut at room temperature using a microtome fitted with a glass knife. Usually 2 μm thick sections are cut. The sections are floated out on water, mounted on glass slides and air dried for 30–60 min before use.

3.3 Cell preparations

Enzyme cytochemistry has proven to be a useful tool for the diagnosis of diseases of blood and bone marrow (for a recent review, see Hayhoe and Quaglino 1988). Furthermore, the analysis of enzyme activities in cultured cells or cells isolated from a tissue such as fibroblasts or hepatocytes can reveal important aspects of cellular metabolism. Enzyme activity can vary greatly as it may be dependent on specific properties of the cells such as the cell line or cell type involved, or the stage of maturation or differentiation. As a consequence, cell preparations may consist of mixed cell populations and biochemical assays performed on homogenates yield results of limited meaning (Hayhoe and Quaglino 1988). This problem is circumvented by applying biochemical assays to purified cell populations (Stuart *et al.* 1975) but cytochemistry is a more appropriate approach. The enzyme activity of individual cells can be detected provided that morphology has been retained during the staining procedure and that the cell type can be recognized afterwards.

Enzyme cytochemical reactions on individual cells are usually performed on cell smears or cytospin preparations after air drying (Section 3.3.1) or fixation (Section 3.3.2) (Stuart *et al.* 1975; Hayhoe and Quaglino 1988).

Again, it should be noted that fixation leads to variable and considerable loss of enzyme activity and should be avoided as far as possible (see Section 3.1). Therefore, the best options for enzyme cytochemistry applied to individual cells are those used for enzyme histochemistry using unfixed cryostat sections. The following particular conditions should be fulfilled for valid cytochemistry on individual cells (Van Noorden *et al.* 1989):

- enzyme activity should not be lost;
- morphology should be retained;
- loss of cells should be prevented;
- cell membranes should be permeable allowing reagents to diffuse into cells.

3.3.1 Unfixed cell preparations

Method 12 describes the preparation method of unfixed cells for enzyme cytochemical analysis. The last three of the above mentioned conditions are fulfilled when fixed cell preparations are used but the first condition usually cannot be met after fixation. However, when unfixed cell preparations are handled according to Method 12 and the enzyme incubations are performed in the presence of polyvinyl alcohol (see Section 4.2), excellent morphology can be kept without loss of enzyme activity. The localization of the final reaction product can be precise enough to recognize organelles within cells in which the enzyme activity is present (Fig. 3.1). Sufficient permeability of cell membranes is normally achieved by drying cell preparations, but an extra freeze–thawing step may enhance permeability. Usually short incubation periods are needed (5–30 min at 37 °C) and only small amounts of medium are required for the reaction (250 µl per slide) (Van Noorden *et al.* 1989). For these reasons, we recommend the use of unfixed cell preparations in combination with the addition of polyvinyl alcohol to the incubation media as the methods of choice for (routine) enzyme cytochemistry using individual cells.

Method 12. *Unfixed cytospin preparations*

1. Centrifuge cell suspension for 10 min at 4 °C (50–2000 *g*, depending on the cell type).
2. Resuspend cells in an Earle's salt solution containing 5 per cent (w/v) serum albumin.
3. Use 50 µl of cell suspension with approximately 1×10^6 cells/ml for cytospin preparations. Stick cells to glass slides using a cytospin centrifuge or special buckets in a standard centrifuge. Centrifuge for 5 min at 400 *g* at room temperature.
4. Dry preparations in a desiccator and store at −25 °C (see Section 3.4.2).

3.3.2 Fixed cell preparations

Individual cells are best fixed in suspension according to Method 13.

Method 13. *Fixation of individual cells*

1. Fix cells in suspension using 1–2 per cent glutaraldehyde and 6 per cent sucrose dissolved in 100 mM cacodylate buffer (pH 7.2) for 10–30 min at 4 °C.
2. Centrifuge cell suspension for 10 min at 4 °C (50–2000 *g*, depending on the type of cell).
3. Rinse cells overnight in a solution of 6 per cent sucrose in 100 mM cacodylate buffer (pH 7.2) at 4 °C.
4. Repeat step 2.
5. Suspend cells in a 7.5 per cent agar solution and let it solidify.
6. Cut and freeze solidified cell suspension as described in Method 4.

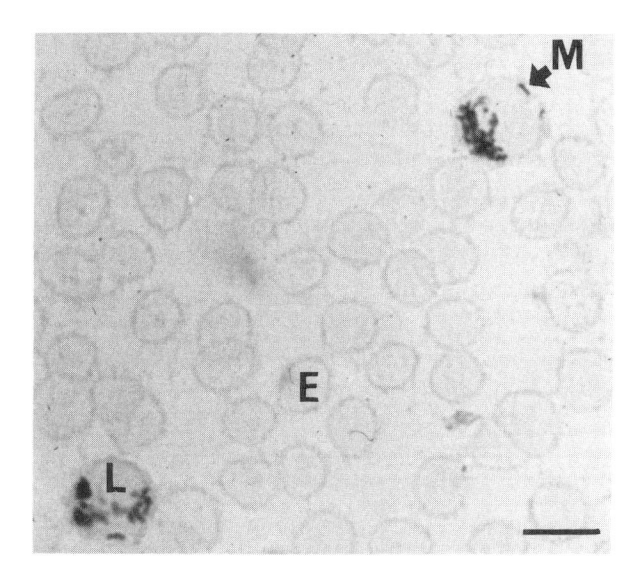

Fig. 3.1. Photomicrograph of unfixed peripheral blood cells incubated for glutamate dehydrogenase activity according to Method 33; cell preparation was according to Method 12. Note that the final reaction product is precipitated exclusively in elongated structures within the cytoplasm, resembling mitochondria. E, erythrocyte; L, leukocyte, possibly a monocyte; M, elongated mitochondrion; bar, 8 μm.

3.4 Storage of cell and tissue preparations

3.4.1 Storage of tissue blocks

3.4.1.1 Storage of frozen tissue blocks

When tissue is snap frozen according to Method 4 it can be stored at −80 °C in a deep-freeze or in a liquid nitrogen specimen storage container for indefinite periods. There is no significant loss of enzyme activity (Henderson et al. 1981) or tissue morphology provided that the tissue is kept in air-tight vials, sealed polyethylene bags or wrapped in aluminium foil, Parafilm or cellophane. Air-tight storage prevents the tissue from drying out. When tissue blocks have to be used for sectioning on different occasions, this can be done by transferring the specimen from the storage container to the cryostat cabinet and vice versa in liquid nitrogen or solid CO_2. The tissue blocks can be kept at −25 °C in the cryostat cabinet for shorter periods during experimental procedures. As long as the temperature of the tissue does not exceed −25 °C, the tissue is not damaged by ice crystals (Frederiks et al. 1992).

3.4.1.2 Storage of glycol methacrylate-embedded tissue

Fixed or freeze-dried tissue that has been mounted in glycol methacrylate according to Sections 3.1.3 and 3.1.4 should be stored at 4 °C or −20 °C (Pretlow *et al.* 1987; Murray *et al.* 1988) although Murray and Ewen (1989) did not find any reduction of enzyme activity when the blocks were kept at room temperature without any special precautions for one year. However, since comparative data are available only for a limited number of enzymes (Murray and Ewen 1989), it is recommended that glycol methacrylate-mounted tissue is kept at −20 °C or no higher than 4 °C in air-tight containers (Pretlow *et al.* 1987).

3.4.2 Storage of cytospin preparations and sections

When cryostat sections or cytospin preparations have been prepared, they should be stored at −25 °C in the cryostat cabinet until they are used for the enzyme incubation. The slides can be kept at this temperature for at least a week without any significant loss of tissue morphology or enzyme activity (Henderson *et al.* 1981; Frederiks *et al.* 1992). This is true of all enzymes described in this Handbook with the exception of xanthine oxidoreductase (Kooij *et al.* 1991).

Data are not available with respect to storage of sections from glycol methacrylate-mounted tissue except that they can be kept for 30–60 min at room temperature. Nevertheless, these sections undoubtedly retain morphology and enzyme activity when stored at −25 °C as is possible for unfixed cryostat sections or cytospin preparations.

References

Chayen, J., Bitensky, L., and Butcher, R. G. (1973). *Practical Histochemistry*. Wiley & Sons, London.

Frederiks, W. M., Ouwerkerk, I. J. M., Bosch, K. S., Marx, F., Kooij, A., and Van Noorden C. J. F. (1992). The effect of storage of cryostat sections on the activity of enzymes. A quantitative histochemical study. *Histochemical Journal*, **24**, in press.

Hayhoe, F. G. J. and Quaglino, D. (1988). *Haematological Cytochemistry*, 2nd edn. Churchill Livingstone, Edinburgh.

Henderson, B., Loveridge, N., Robertson, W. R., and Smith, M. T. (1981). The influence on enzyme activity of storage of tissue blocks at −70°C. *Histochemistry*, **72**, 545–50.

Hulstaert, C. E., Kalicharan, D., and Hardonk, M. J. (1989). Cytochemical demonstration of phosphatases with the cerium method. In *Techniques in Diagnostic Pathology*, Vol. 1 (ed. G. R. Bullock), pp. 133–50. Academic Press, New York.

Kooij, A., Frederiks, W. M., Gossrau, R., and Van Noorden, C. J. F. (1991). Localization of xanthine oxidoreductase activity using the tissue protectant polyvinyl alcohol and final electron acceptor tetranitro BT. *Journal of Histochemistry and Cytochemistry*, **39**, 87–93.

Lojda, Z., Gossrau, R., and Schiebler, T. H. (1979). *Enzyme Histochemical Methods. A Laboratory Manual.* Springer Verlag, New York.

Murray, G. I. and Ewen, S. W. B. (1989). A new approach to enzyme histochemical analysis of biopsy specimens. *Journal of Clinical Pathology*, **42**, 767–71.

Murray, G. I. and Ewen, S. W. B. (1990). Enzyme histochemistry on freeze-substituted glycol methacrylate-embedded tissue. *Journal of Histochemistry and Cytochemistry*, **38**, 95–101.

Murray, G. I., Burke, M. D. and Ewen, S. W. B. (1988). Dehydrogenase enzyme histochemistry on freeze-dried or fixed resin-embedded tissue. *Histochemical Journal*, **20**, 491–8.

Pretlow, T. P., Grane, R. N., Goehring, P. L., Lapinsky, A. S., and Pretlow II, T. G. (1987). Examination of enzyme-altered foci with gamma-glutamyl transpeptidase, aldehyde dehydrogenase, glucose-6-phosphate dehydrogenase, and other markers in methacrylate-embedded liver. *Laboratory Investigation*, **56**, 96–100.

Stuart, J., Gordon, P. A., and Lee, T. R. (1975). Enzyme cytochemistry of blood and marrow cells. *Histochemical Journal*, **7**, 471–87.

Van Noorden, C. J. F. and Hulstaert, C. E. (1991). Electron microscopical enzyme histochemistry. In *Electron Microscopy of Tissues, Cells, and Organelles* (ed. J. R. Harris), pp. 125–49. Oxford University Press, Oxford.

Van Noorden, C. J. F., Vogels, I. M. C., and Van Wering, E. R. (1989). Enzyme cytochemistry of unfixed leukocytes and bone marrow cells using polyvinyl alcohol for the diagnosis of leukemia. *Histochemistry*, **92**, 313–18.

4 Enzyme histochemical methods

The incubation procedures for the demonstration of enzyme activities can be performed in different ways and the choice of procedure depends on the properties of the enzyme under study and/or the principle of the enzyme histochemical method. Three incubation conditions will be discussed: aqueous incubation media, incubation media containing polyvinyl alcohol, and the semipermeable membrane technique using gelled incubation media.

4.1 Aqueous media

Aqueous incubation media can only be used when the enzyme under study cannot diffuse from the section or cell into the medium. This is the case for instance when fixation has taken place. However, it is clear from the previous chapter that almost all enzymes are inactivated or denatured during fixation, so fixation should be avoided as much as possible. When tissues are embedded in glycol methacrylate (see Sections 3.1.3 and 3.1.4), the enzymes are physically stabilized in the resin and so they may be properly demonstrated with aqueous incubation media. Aqueous media can also be used successfully to localize the activity of tightly bound enzymes such as succinate dehydrogenase, which is present in the inner membrane of mitochondria.

4.2 Media containing polyvinyl alcohol

In 1965, Altman and Chayen introduced inert polymers such as polyvinyl alcohol as tissue protectants in enzyme histochemistry. Addition of polyvinyl alcohol to an incubation medium enables many types of enzymes to be localized in unfixed cryostat sections or cell preparations without loss of the enzymes and other large molecules from the tissue. Inactivation of enzymes by fixation can be avoided and all enzyme activity present in a tissue specimen can be demonstrated, but tissue integrity is still preserved.

When polyvinyl alcohol is dissolved in water or buffer, the solution becomes viscous, but it is not this viscosity that has the protective effect on sections. Rather, a tissue is protected by the principle of the excluded volume, i.e. solvent water molecules are trapped by the polymer rods so that less solvent water is available for diffusion of molecules from tissue sections.

A number of alternatives such as polyethylene glycol, polyvinyl pyrrolidone, Polypep and others have been tested for use in enzyme histochemistry in order to circumvent the high viscosity of solutions containing polyvinyl alcohol. However, these alternatives have never become as widely used in enzyme histochemistry as polyvinyl alcohol for various reasons (for review, see Van Noorden and Vogels 1989a). The addition of polyvinyl alcohol to the incubation medium may improve the precision of localization for reasons other than restriction of diffusion. Low concentrations of polyvinyl alcohol allow colloidal dispersions of sufficiently high concentrations of compounds which are not very water soluble, such as tetranitro BT or diaminobenzidine, to be maintained in incubation media (Kugler et al. 1988b).

A consequence of the addition of polyvinyl alcohol to incubation media is that most reagents have to be added in considerably higher concentrations than in aqueous media, in order to obtain similar reaction rates. This may be due to two factors; firstly, diffusion of small molecules may be hampered by the presence of polyvinyl alcohol. Secondly, protection of unfixed cryostat sections by polyvinyl alcohol implies that enzymes exert activity in their own cellular environment (see Section 2.3). Originally polyvinyl alcohol was used to prevent diffusion of so-called 'soluble' enzymes from tissue sections or cell preparations. However, later on it became clear that localization of tightly bound enzymes can be improved considerably by the use of polyvinyl alcohol as well (Van Noorden et al. 1989a). This is most probably due either to decreasing diffusion of intermediate reaction products from the enzymic sites or to the build up of sufficient concentrations of trapping reagents at the enzymic sites due to the presence of polyvinyl alcohol in the medium, thus giving rise to a sharper localization (Fig. 3.1).

Not all grades of polyvinyl alcohol available from the various companies yield reliable results. Only polyvinyl alcohols with a number average molecular weight of 20 000–40 000 appeared to be suitable as tissue protectants. Moreover, it was concluded from qualitative and quantitative studies that 18 per cent (w/v) polyvinyl alcohol from Sigma (Sigma Chemical Company, St Louis, MO, USA) with a number average molecular weight of 40 000 (see also the note to Method 14) is the tissue protectant of choice for enzyme histochemical reactions in general. The preparation of media containing polyvinyl alcohol is covered in Method 14.

Method 14. *Preparation of polyvinyl alcohol-containing medium*

1. Dissolve 18 g polyvinyl alcohol (Sigma, St Louis, MO, USA; number average mol. wt, 40 000*) in 100 ml buffer, stirring and heating in a water bath until a clear solution is obtained.
2. Store the clear solution at 60 °C in air-tight vials.

3. Cool a desired volume of the solution to 37 °C before incubation.
4. Prepare incubation medium by adding to the polyvinyl alcohol-containing buffer solution at 37 °C the appropriate substrates, coenzyme, and other compounds in dissolved form as concentrated stock solutions. Add to 1 ml buffer, 10 µl of concentrated stock solution which contains 100 times the final concentration.
5. Mix the compounds thoroughly with the buffer solution using a spatula after each addition because of the viscosity of the medium.

* This type of polyvinyl alcohol from Sigma has a number average molecular weight of 40 000 and a weight average molecular weight of 70 000–100 000 (Dr A. O'Connor, Sigma, personal communication). In the past, the number average molecular weight has usually been used to express the molecular weights of polymers, whereas now the weight average molecular weight is preferred.

There are some specific conditions which do not allow polyvinyl alcohol to be used in incubation media (see Section 4.3). In these cases, aqueous incubation media or the application of the semipermeable membrane technique are to be preferred (Section 4.1 and 4.3).

4.3 Semipermeable membrane technique

Diffusion of enzymes from unfixed sections or cell preparations can be prevented by addition of polyvinyl alcohol to incubation media (see Section 4.2), but diffusion can also be prevented by interposing a semipermeable membrane between the section and the incubation medium as introduced by McMillan (1967). A combination of the semipermeable membrane technique with a gelled incubation medium has been developed by Meijer (1972). This technique can be used for the demonstration of activities of hydrolases, oxidoreductases, transferases, and isomerases (Meijer 1980). It is, however, rather time-consuming and therefore we recommend the use of a polyvinyl alcohol-containing incubation medium as an easier procedure with similar advantages. When the polyvinyl alcohol procedure fails, the semipermeable membrane technique may be a useful alternative. When the final reaction product is soluble in hot water or buffer, which is necessary to rinse off the viscous polyvinyl alcohol-containing incubation medium, the semipermeable membrane technique combined with a gelled incubation medium is the method of choice. In this procedure the gelled incubation medium can be easily removed with a spatula without using an aqueous solvent. The technique is recommended for the demonstration of glycogen phosphorylase activity (see Method 46).

The combination of the semipermeable membrane technique and a gelled incubation medium is also recommended when cerium ions are used as the capturing reagent. These ions probably form a complex with polyvinyl alcohol polymers, so that free ions are not available for the capturing reaction (Jonges *et al.* 1990; Frederiks *et al.* 1990). The semipermeable membrane technique has also been successful for the demonstration of the activity of a peroxisomal enzyme, D-amino acid oxidase (Patel *et al.* 1991).

The principles of the semipermeable membrane technique are based on the preparation of an incubation vessel by stretching a piece of dialysis tubing over a Perspex ring and fastening it with elastic bands. This incubation vessel is then filled with a warm agar solution containing all reagents needed for the enzyme reaction. After the medium has solidified, the cryostat section is fixed to the membrane and the incubation starts (see Method 15).

Method 15. *Preparation of membranes and media for the semipermeable membrane technique*

1. Soak dialysis membranes (type 36/32", wall thickness 7.2×10^{-4} inch, average pore radius 24 Å; manufactured by Visking Co., Chicago, USA) in 1 per cent (w/v) EDTA by boiling for one hour to remove impurities and trace metals. Rinse three times in distilled water and remove EDTA by boiling for another hour in distilled water. Rinse three times in distilled water and store in 98 per cent ethanol at 4 °C until required. Dialysis tubing manufactured by companies other than Visking does not usually give such satisfactory results.
2. Close one end of rings (height, 2.5 cm; diameter, 3 cm) made out of Perspex tubes with a piece of semipermeable membrane, 5 cm square, which has been rinsed in distilled water. Stretch moist tubing and secure with elastic bands.
3. Prepare a 3 per cent (w/v) agar (Nobel grade, Difco Laboratories, Detroit, USA) solution in buffer, by heating in a water bath. Mix one agar solution volume at about 60 °C with an equal volume of a solution containing all the constituents in concentrations twice as high as the desired final concentrations.
4. Pour 2 ml portions of the mixture rapidly into the membrane-covered incubation vessels and allow to solidify at room temperature. The solid gel should be at least 0.5 cm thick. If necessary, store the prepared vessels overnight in a closed jar at 4 °C, but preferably use them immediately.
5. Dry the upper surface of the membranes with filter paper just before use.
6. Mount cryostat-cut (5–10 μm thick) sections of unfixed tissue directly from the cryostat knife onto the membranes.
7. Incubate the sections for the required time at 37 °C in the dark in a dry environment with the section on top of the vessel.
8. Remove the gelled incubation medium with a spatula and wipe off any adhering gel.
9. Incubate sections with adhering membranes in a second step when necessary (see respective Methods).
10. Fix the sections and their adhering membranes with formaldehyde vapour for 5 mins at room temperature. (This step is optional.)
11. Humidify the sections with adhering membranes in distilled water.
12. Cut out the sections and underlying membranes with scissors.

13. Mount the sections in glycerine–gelatin on an object slide (membrane-mounted sections cannot be readily dehydrated and mounted in xylene-containing permanent mountants: the membranes harden and curl).

4.4 Enzyme histochemical procedures

On the basis of the principles of enzyme histochemical methods as described in Section 2.6, a selection of procedures is described in this chapter. The procedures have been selected from those described in the literature on the basis of the opinion of the authors that the method of choice has been tested with respect to most or all aspects of validity as discussed in Section 2.2. For each enzyme, the function is given as well as the tissues in which the activity has been demonstrated under optimal conditions. It should be realized that the medium may have to be adapted to demonstrate the activity in other tissues because kinetic properties of an enzyme may differ in various tissues. The subcellular localization, pH optimum, substrates, other necessary constituents, and (specific) inhibitors of the enzymes are also presented.

When procedures are described for fixed cryostat sections or cryostat sections from fixed tissues using aqueous incubation media, the method may be improved by application of unfixed cryostat sections and addition of polyvinyl alcohol to the incubation medium.

4.4.1 Metal salt capture methods

This rather old principle for the demonstration of activities of phosphatases, sulfatases, nucleotidases, and cholinesterases is used less and less frequently due to methodological problems and localization artefacts. Two phosphatases are still detected using lead ions: 5′-nucleotidase (Method 16) and adenylate cyclase (Method 18). Lead ions are replaced by strontium ions for the demonstration of ATPase activity (Method 17). Lead ions inhibit ATPase considerably, cause non-enzymatic hydrolysis of ATP, and form insoluble complexes with ATP (Ernst 1972). Lead is applied as a capturing reagent for phosphate produced by purine nucleoside phosphorylase (Method 19) and ornithine carbamoyl transferase (Method 20), for carbon dioxide produced by ornithine decarboxylase (Method 21), for oxaloacetate produced by aspartate aminotransferase (Method 22), and for sulfate produced by arylsulfatase (Method 23).

5′-Nucleotidase (EC 3.1.3.5)

Function: degradation of nucleic acids and transport of nucleotides through cell membranes

Tissues: liver, kidney, small intestine, heart muscle, nerve tissue, testis, arterial wall
Subcellular localization: plasma membranes, lysosomes
pH optimum: 7.2
Validity: quantiative (rat liver; Frederiks and Marx 1988)
Substrate: AMP
Other constituents: magnesium chloride
Inhibitors: sodium fluroide, EDTA.

Method 16. *Lead salt capture method for 5'-nucleotidase activity (Frederiks and Marx 1988)*

1. Prepare 100 ml of 0.1 M Tris–maleate buffer, pH 7.2, containing 18 per cent polyvinyl alcohol according to Method 14.
2. Add to 1 ml polyvinyl alcohol-containing medium 20 μl of a stock solution of lead nitrate (119 mg/ml distilled water; final concentration 7.2 mM) at 37 °C. Mix immediately and thoroughly to prevent precipitation.
3. Then, add in order, 10 μl of a stock solution of magnesium chloride (203 mg/ml buffer; final concentration 10 mM) and 10 μl of a stock solution of AMP (250 mg/ml buffer; final concentration 5 mM). The medium is clear for only 15–30 min.
4. Use air-dried, unfixed cryostat sections, 5–10 μm thick.
5. Pour the medium (200–500 μl per section) on to each section.
6. Incubate at 37 °C in the dark for 5–30 min.
7. After incubation, stop reaction immediately by washing sections in distilled water at 60 °C.
8. Incubate the sections for 1 min in a solution of 1 per cent (v/v) ammonium sulfide in distilled water in order to convert all precipitated lead phosphate into black lead sulfide.
9. Rinse the sections four times in distilled water and postfix in 4 per cent (v/v) formaldehyde in distilled water for 10 min at room temperature.
10. Rinse in distilled water and mount in glycerine–gelatin.
11. Control incubations should be carried out (i) in the absence of AMP; (ii) in the presence of β-glycerophosphate instead of AMP; (iii) in the presence of AMP and 10 mM tetramisole, an inhibitor of alkaline phosphatase activity; and (iv) in the presence of AMP and 50 mM sodium fluoride or 50 mM EDTA, aspecific inhibitors of 5'-nucleotidase.

ATPases (EC 3.6.1.3)

ATPases share the ability to catalyse the splitting of ATP into ADP and orthophosphate or the synthesis of ATP from ADP and phosphate. Various ATPases are responsible for this reaction but they differ in their intracellular localization, their optimum pH, and their sensitivity to activators or inhibitors. A reliable method has been described for Na$^+$, K$^+$-ATPase; however, it seems that the calcium-activated myosin ATPase can be demonstrated with the same method when using calcium ions instead of sodium, potassium and magnesium ions as activators (for review, see Johnson 1991).

Na$^+$, K$^+$-ATPase

Function:	sodium pump
Tissues:	kidney, liver, avian salt gland
Subcellular localization:	plasma membrane
pH optimum:	7.2–7.4
Validity:	quantitative (mouse kidney; Firth 1987)
Substrate:	p-nitrophenyl phosphate
Other constituents:	Na$^+$, K$^+$, Mg^{2+}
Inhibitors:	ouabain, sodium fluoride, Pb^{2+}, Ca^{2+}, blockers of SH-groups.

Ca^{2+}-myosin ATPase

Function:	involved in muscular contraction
Tissues:	heart, skeletal muscle
Subcellular localization:	A-band of sarcomeres
pH optimum:	7.2–7.4
Validity:	unknown
Substrate:	p-nitrophenyl phosphate
Other constituents:	Ca^{2+}
Inhibitors:	Pb^{2+}, blockers of SH-groups.

Method 17. *Indirect p-nitrophenyl phosphate method for Na$^+$, K$^+$-ATPase and Ca^{2+}-myosin ATPase (Firth 1987)*

---

1. Prepare the incubation medium by dissolving in 100 ml distilled water: 620 mg 2-amino-2-methyl-1-propanol (final concentration 70 mM), 160 mg SrCl$_2$ (final concentration 1 mM), 220 mg KCl (final concentration 30 mM), 48 mg MgCl$_2$ (final concentration 5 mM), (for Ca^{2+}-myosin ATPase: 78 mg CaCl$_2$ (final concentration 7 mM) instead of KCl and MgCl$_2$), 25 mg dimethylsulfoxide (final concentration 3.2 mM), 190 mg p-nitrophenyl phosphate disodium salt (final concentration 5 mM) and 37 mg L-p-bromotetramisole oxalate (final concentration 1 mM). Adjust the pH to 9.0 with HCl.
2. Use air-dried unfixed cryostat sections, 5–10 µm thick.
3. Incubate sections for 15–90 min at 37 °C.
4. Rinse sections twice in distilled water.
5. Incubate the sections for 4 min in a solution of 2 per cent (w/v) cobalt chloride in distilled water in order to convert strontium phosphate into cobalt phosphate.

6. Rinse the sections four times in distilled water and incubate for 1 min in a solution of 1 per cent (v/v) ammonium sulfide in distilled water in order to convert cobalt phosphate into cobalt sulfide (black).
7. Rinse the sections four times in distilled water and postfix for 10 min in 4 per cent (v/v) formaldehyde in distilled water at room temperature.
8. Rinse in distilled water and mount in glycerine–gelatin.
9. Control incubations should be carried out (i) in the absence of substrate; and (ii) in the presence of 5 mM ouabain, an inhibitor of Na^+, K^+-ATPase.

Adenylate cyclase (EC 4.6.1.1)

Function:	production of cAMP which modulates the intracellular level of Ca^{2+} and serves as second messenger
Tissues:	heart, liver, kidney, endocrine pancreas, thymocytes salivary gland, nervous tissue
Subcellular localization:	plasma membranes
pH optimum:	7.4
Validity:	qualitative (rat liver; Mayer et al. 1985)
Substrates:	adenylate-(β, γ-methylene)-diphosphate
Other constituents:	Mg^{2+}
Inhibitor:	alloxan.

Method 18. *Lead salt capture method for adenylate cyclase activity (Mayer et al. 1985)*

1. Use cryostat sections of 5–10 μm thickness from tissues fixed according to Methods 5–7.
2. Prepare the incubation medium by dissolving 72 mg theophylline (final concentration 4 mM; inhibitor of cAMP phosphodiesterase), 80 mg lead citrate (final concentration 2 mM), 59 mg magnesium nitrate (final concentration 4 mM), 41 mg tetramisole (final concentration 2 mM; inhibitor of alkaline phosphatase), 84 mg sodium fluoride (final concentration 20 mM; inhibitor of 5′-nucleotidase) and 43 mg AMP-P(CH$_2$)P (final concentration 1 mM) in 100 ml 0.1 M HEPES buffer, pH 7.4 containing 5 per cent sucrose.
3. Incubate sections for 15–60 min at 37 °C.
4. Rinse sections three times in distilled water.
5. Incubate sections for 1 min in a solution of 1 per cent (v/v) ammonium sulfide in distilled water in order to convert lead phosphate into black lead sulfide.
6. Rinse sections four times in distilled water.
7. Mount in glycerine–gelatin.
8. Control incubations should be carried out (i) in the absence of substrate; (ii) by replacing AMP-P(CH$_2$)P with 3.0 mM sodium β-glycerophosphate; (iii) by adding 10 mM alloxan, a specific inhibitor of adenylate cyclase; (iv) by adding 0.05 mg/ml glucagon, an activator of adenylate cyclase; and (v) by adding 0.125 mg/ml cholera toxin, an activator of adenylate cyclase.

Purine nucleoside phosphorylase (EC 2.4.2.1)
Function:	breakdown of inosine
Tissues:	endothelial cells, fibroblasts, glial cells, neutrophils, basophils, monocytes, platelets
Subcellular localization:	cytoplasm
pH optimum:	7.4
Validity:	qualitative (glial cells; Van Reempts *et al.* 1988)
Substrates:	ribose-1-phosphate and hypoxanthine
Other constituents:	–
Inhibitor:	*p*-chloromercuribenzoic acid, *N*-ethylmaleimide.

Method 19. *Lead salt capture method for purine nucleoside phosphorylase activity (Van Reempts et al. 1988)*

1. Fix tissues or organs according to Methods 5–7.
2. Fix the tissue via immersion in the same fixative for an additional 20 min.
3. Prepare 100 µm thick vibratome sections.
4. Fix the sections via immersion in the above mentioned fixative for 35 min at 4 °C.
5. Rinse in ice-cold 60 mM Tris–maleate buffer, pH 7.4, containing 220 mM sucrose.
6. Freeze sections in vials in liquid nitrogen.
7. Prepare incubation medium by dissolving in 100 ml 60 mM Tris–maleate buffer (pH 7.4), 100 mg lead nitrate (final concentration 3 mM), 92 mg ribose-1-phosphate (final concentration 4 mM), 14 mg hypoxanthine (final concentration 10 mM) and 7.53 g sucrose (final concentration 220 mM).
8. Incubate sections for 10–30 min at 37 °C.
9. Rinse sections in 220 mM sucrose in distilled water.
10. Treat sections for 1 min in a solution of 1% (v/v) ammonium sulfide in distilled water.
11. Rinse sections in distilled water.
12. Mount in glycerine–gelatin.
13. Control incubations should be carried out (i) by replacing ribose-1-phosphate with ribose-5-phosphate; (ii) in the absence of hypoxanthine; (iii) in the absence of ribose-1-phosphate and hypoxanthine; and (iv) in the presence of 1 mM *p*-chloromercuribenzoic acid.

Ornithine carbamoyltransferase (EC 2.1.3.3)
Function:	involved in urea synthesis
Tissues:	liver, small intestine
Subcellular localization:	mitochondrial matrix
pH optimum:	7.7
Validity:	qualitative (liver; Holzgreve *et al.* 1985)
Substrates:	carbamoyl phosphate and L-ornithine
Other constituents:	–
Inhibitor:	L-leucine.

Method 20. *Lead salt capture method for ornithine carbamoyltransferase (Holzgreve et al. 1985)*

1. Fix small pieces of tissue according to Method 7.
2. Rinse the tissue in 0.05 M cacodylate buffer, pH 7.4.
3. Freeze the tissue in vials in liquid nitrogen.
4. Cut cryostat sections and mount on slides.
5. Prepare 0.025 M triethanolamine buffer (pH 7.7).
6. Add to 9 ml buffer 1 ml of a 1 per cent lead nitrate solution in distilled water (final concentration 3 mM).
7. Dissolve in the mixture 3 mg carbamoylphosphate (lithium salt) (final concentration 2 mM), 5 mg L-ornithine (final concentration 3.7 mM) and 0.8 g sucrose (final concentration 0.23 M).
8. Incubate sections for 5–15 min in the incubation medium at 37 °C.
9. Rinse in 0.025 M triethanolamine buffer (pH 7.7).
10. Rinse three times in distilled water.
11. Treat sections for 1 min in a solution of 1 per cent (v/v) ammonium sulfide in distilled water.
12. Rinse sections in distilled water.
13. Mount in glycerine–gelatin.
14. Control incubations should be carried out (i) in the absence of ornithine; (ii) in the absence of ornithine and carbamoyl phosphate; and (iii) in the presence of L-leucine, a specific inhibitor of ornithine carbamoyltransferase.

Ornithine decarboxylase (EC 4.1.1.17)

Function:	key enzyme in biosynthesis of polyamines, which play a role in cellular growth and development
Tissues:	liver, kidney, prostate, brain
Subcellular localization:	cytoplasm
pH optimum:	7.0–7.8
Validity:	quantitative (mouse kidney; Dodds *et al.* 1990)
Substrate:	L-ornithine
Other constituents:	pyridoxal phosphate, dithiothreitol
Inhibitor:	α-difluoromethyl ornithine.

Method 21. *Lead salt capture method for ornithine decarboxylase (Dodds et al. 1990)*

1. Prepare all solutions in CO_2-free water.
2. Dissolve 26 g polyvinyl alcohol in 100 ml 0.2 M triethanolamine buffer (pH 7.0) according to Method 14.
3. Prepare the incubation medium by adding to 0.7 ml polyvinyl alcohol-containing buffer 0.3 ml lead hydroxyisobutyrate solution in distilled water (stock 102 mg/ml; final concentration 0.1 M).

4. Add to the mixture 10 µl L-ornithine (stock 340 mg/ml; final concentration 11.4 mM), 10 µl pyridoxal phosphate (stock 62 mg/ml; final concentration 0.38 mM), 10 µl D,L-dithiothreitol (stock 8 mg/ml; final concentration 0.6 mM) and 10 µl L-*p*-bromotetramisole oxalate (stock 15 mg/ml; final concentration 0.3 mM).
5. If necessary, adjust the pH to pH 7.0.
6. Use air-dried unfixed cryostat sections, 5–10 µm thick.
7. Pour the medium (200–500 µl) onto sections.
8. Incubate at 37 °C in the dark for 10–30 min.
9. Wash sections in CO_2-free water at 60 °C.
10. Incubate the sections for 1 min in a solution of 1 per cent (v/v) ammonium sulfide in distilled water.
11. Rinse the sections four times in distilled water.
12. Mount in glycerine–gelatin.
13. Control incubations should be carried out (i) in the absence of L-ornithine; and (ii) in the presence of 40 mM α-difluoromethyl ornithine, a specific irreversible inhibitor of ornithine decarboxylase.

Aspartate aminotransferase (EC 2.6.1.1)

Function: a central role in amino acid metabolism
Tissues: heart, skeletal muscle, liver, kidney
Subcellular localization: mitochondria, cytoplasm
pH optimum: 7.2
Validity: quantitative (liver; Papadimitriou and Van Duijn 1970)
Substrates: α-ketoglutarate and L-aspartate
Other constituents: –
Inhibitor: L-glutamic acid.

Method 22. *Lead salt capture method for aspartate aminotransferase (Papadimitriou and Van Duijn 1970)*

1. Fix tissues via perfusion according to Methods 5 and 6.
2. Fix small pieces of tissue (0.5 cm^3) for 10 min in 0.37 per cent formaldehyde in 100 ml 0.05 M imidazole buffer (pH 7.2), containing 1 g sucrose (final concentration 29 mM) and 117 mg α-ketoglutaric acid (final concentration 8 mM).
3. Wash for 1 h in cold 0.25 M sucrose in 0.05 M imidazole buffer at 4 °C.
4. Freeze in liquid nitrogen.
5. Cut cryostat sections with a thickness of 5–10 µm.
6. Fix for 2 min in 3.7 per cent formaldehyde in 0.05 M imidazole buffer (pH 7.2) at 4 °C.
7. Wash in 0.05 M imidazole buffer (pH 7.2).
8. Prepare the incubation medium by dissolving in 100 ml 0.05 M imidazole buffer (pH 7.2) 58 mg α-ketoglutaric acid (final concentration 4 mM), 266 mg L-aspartic acid (final concentration 20 mM) and 200 mg lead nitrate (final concentration 6 mM).

9. Incubate for 20 min at 22 °C.
10. Wash in 100 ml 0.05 M imidazole buffer (pH 7.2) containing 20 mM L-aspartic acid.
11. Treat for 1 min with a solution of 1 per cent (v/v) ammonium sulfide in distilled water.
12. Rinse four times in distilled water.
13. Mount in glycerine–gelatin.
14. Control incubations should be carried out (i) in the absence of α-ketoglutaric acid and L-aspartic acid; (ii) in the presence of D-aspartic acid instead of L-aspartic acid; and (iii) in the presence of 10 mM L-glutamic acid, a specific inhibitor of aspartate amino-transferase.

Arylsulfatase (EC 3.1.6.1)

Function:	unknown
Tissues:	kidney, liver, brain, uterus
Subcellular localization:	A and B isoenzymes in lysosomes, C isoenzyme in endoplasmic reticulum
pH optimum:	5.5
Validity:	qualitative (rat kidney; Hopsu-Havu et al. 1967)
Substrate:	nitrocatechol sulfate
Other constituents:	–
Inhibitor:	sodium sulfite.

Method 23. *Metal salt capture method for arylsulfatase (Hopsu-Havu et al. 1967)*

1. Fix tissue blocks according to Method 7.
2. Rinse in 0.1 M cacodylate buffer (pH 7.3).
3. Freeze in liquid nitrogen.
4. Cut cryostat sections with a thickness of 5–10 μm.
5. Prepare the incubation medium by dissolving 160 mg *p*-nitrocatechol sulfate (final concentration 25 mM) and 320 mg lead nitrate (final concentration 48 mM) in 20 ml 0.1 M sodium acetate buffer (pH 5.5).
6. Incubate for 15–60 min at 37 °C.
7. Rinse in distilled water.
8. Treat for 1 min with 1 per cent (v/v) ammonium sulfide in distilled water.
9. Rinse four times in distilled water.
10. Mount in glycerine–gelatin.
11. Control incubations should be carried out (i) in the absence of *p*-nitrocatechol sulfate; and (ii) in the presence of sodium sulfite, a specific inhibitor of arylsulfatase.

4.4.2 Cerium salt capture methods

Cerium salt methods are promising for the detection of phosphatases and oxidases. At this moment a reliable and quantiative method has been published for glucose-6-phosphatase based on the use of hydrogen peroxide and diaminobenzidine (Jonges *et al.* 1990). The principles of this method have also been described for the demonstration of alkaline phosphatase and ATPase (Van Goor *et al.* 1989; Van Goor and Hardonk 1990). However, these methods have not been optimized and tested for quantification. Moreover, a reliable indoxyl-tetrazolium salt method is available for alkaline phosphatase (Van Noorden and Jonges 1987). Therefore, we do not recommend, at this time, that the cerium salt capture methods are used for phosphatases other than glucose-6-phosphatase.

Glucose-6-phosphatase (EC 3.1.3.9)

Function: key enzyme in the production of glucose
Tissues: liver, kidney
Subcellular localization: nuclear envelope, endoplasmic reticulum
pH optimum: 6.5
Validity: quantitative (rat liver; Jonges *et al.* 1990)
Substrate: glucose-6-phosphate
Other constituents: –
Inhibitor: citric acid.

Method 24. *Cerium-diaminobenzidine-hydrogen peroxide procedure for glucose-6-phosphatase (Jonges et al. 1990)*

1. Use air-dried cryostat sections with a thickness of 5–10 μm.
2. Prepare 0.05 M Tris–maleate buffer (pH 6.5).
3. Prepare the incubation medium by dissolving 123 mg cerium chloride (final concentration 5 mM) and 260 mg glucose-6-phosphate (final concentration 10 mM) in 100 ml 0.05 M Tris–maleate buffer (pH 6.5).
4. Incubate for 5 min at room temperature.
5. Rinse in 0.05 M Tris–maleate buffer (pH 8.0).
6. Prepare the second step medium by dissolving 360 mg diaminobenzidine (final concentration 10 mM) and 65 mg sodium azide (final concentration 10 mM) in 100 ml 0.05 M Tris–maleate buffer, pH 8.0. Add 13 ml hydrogen peroxide to the mixture (stock 30 per cent; final concentration 4 per cent) immediately before use.
7. Incubate for 10 min at room temperature in the second step medium.
8. Rinse in distilled water.
9. Mount in glycerine–gelatin.
10. Control incubations should be carried out (i) in the absence of glucose-6-phosphate; and (ii) in the presence of citric acid, a specific inhibitor of glucose-6-phosphatase.

Oxidases

Briggs *et al.* (1975) introduced a cerium procedure for detection of NADH oxidase at the electron microscopical level. Since then, various peroxisomal oxidases have been demonstrated with this technique. The procedure developed by Angermueller and Fahimi (1988), based on the polymerization of diaminobenzidine by the conversion of cerium perhydroxide, was a breakthrough for the demonstration of oxidases with light microscopy. The colour of polymerized diaminobenzidine can be intensified by adding cobalt chloride (Angermueller and Fahimi 1988) and small amounts of hydrogen peroxide (Gossrau *et al.* 1989) to the visualization medium. This principle enables the detection of peroxisomal oxidases, such as D-amino acid oxidase, α-hydroxy acid oxidase, urate oxidase, and xanthine oxidase. Recently, the technique was optimized for the demonstration of D-amino acid oxidase in rat liver using a semipermeable membrane (Patel *et al.* 1991), but the enzyme has also been demonstrated reliably when an optimized incubation medium is used without applying a semipermeable membrane (Frederiks *et al.* 1992a). Peroxisomal oxidases seem to be bound rather firmly and so aqueous techniques are recommended for the demonstration of these enzymes. Xanthine oxidase may be an exception because this enzyme is localized not only in peroxisomes but also in the cytoplasm. Moreover, a reliable method with the use of a tetrazolium salt and polyvinyl alcohol is available for the demonstration of the activity of xanthine oxidoreductase (both dehydrogenase and oxidase activity) (Kooij *et al.* 1991).

The enzyme for which cerium salt capture methods were originally introduced, i.e. NADH oxidase in activated leukocytes, has not yet been demonstrated at the light microscopical level in these cells. The method has been applied for the demonstration of NAD(P)H oxidase in rat liver (Mizukami *et al.* 1983) but these results do not seem to be reliable due to interference of phosphatases in the reaction (Gossrau *et al.* 1990).

D-amino acid oxidase (EC 1.4.3.3)

Function:	role in regulation of cellular metabolism by the production of oxalate
Tissues:	kidney, liver, brain
Subcellular localization:	peroxisomes
pH optimum:	7.8
Validity:	quantitative (rat liver; Frederiks *et al.* 1992a)
Substrate:	D-proline
Other constituents:	sodium azide
Inhibitors:	D,L-β-hydroxybutyrate, propargylglycine.

Method 25. *Cerium-diaminobenzidine-cobalt-hydrogen peroxide procedure for D-amino acid oxidase (Frederiks et al. 1992a)*

1. Use air-dried cryostat sections.
2. Prepare 0.3 M Tris–maleate buffer (pH 7.8).
3. Prepare incubation medium by heating the buffer to 70 °C and adding to 100 ml buffer 3 ml cerium chloride also heated to 70 °C (371 mg/ml distilled water; final concentration 30 mM). Cool down to 37 °C and add 2 ml sodium azide (325 mg/ml distilled water; final concentration 100 mM) and 2 ml D-proline (115 mg/ml distilled water; final concentration 20 mM).
4. Incubate for 30 min at 37 °C.
5. Rinse three times with distilled water.
6. Prepare the visualization medium by dissolving 50 mg diaminobenzidine in 100 ml 0.05 M sodium acetate buffer (pH 5.3; final concentration 1.4 mM) and by adding to 100 ml buffer 2 ml sodium azide (325 mg/ml distilled water; final concentration 100 mM), 1 g cobalt chloride (final concentration 42 mM) and, immediately before use, 7 µl hydrogen peroxide (stock 30 per cent; final concentration 0.6 mM).
7. Incubate for 30 min at 37 °C in the visualization medium.
8. Rinse in distilled water.
9. Mount in glycerine–gelatin.
10. Store sections in the dark at 4 °C.
11. Control incubations should be carried out (i) in the absence of D-proline; and (ii) in the presence of 200 mM D,L-β-hydroxybutyrate, a specific inhibitor of D-amino acid oxidase.

α-Hydroxy acid oxidase (EC 1.1.3.15)

Function:	oxidation of short chain aliphatic, long chain aliphatic and long chain aromatic hydroxy acids
Tissues:	kidney, liver
Subcellular localization:	peroxisomes
pH optimum:	7.8
Validity:	qualitative (rat liver and kidney; Angermueller 1989)
Substrates:	the liver isozyme, A, reacts with glycolic acid; the kidney isozyme, B, reacts with D,L-α-hydroxybutyric acid
Other constituents:	sodium azide
Inhibitor:	phenyl acetic acid.

Method 26. *Cerium-diaminobenzidine-cobalt-hydrogen peroxide procedure for α-hydroxy acid oxidase (Angermueller et al. 1986; Frederiks et al. 1992a)*

1. Use air-dried cryostat sections.
2. Prepare 0.3 M Tris–maleate buffer (pH 7.8).

3. Prepare the incubation medium by heating the buffer to 70 °C and by adding to 100 ml buffer 3 ml cerium chloride also heated to 70 °C (371 mg/ml distilled water; final concentration 30 mM). Cool down to 37 °C and add 2 ml sodium azide (325 mg/ml distilled water; final concentration 100 mM) and 1 ml D,L-α-hydroxybutyric acid (252 mg/ml distilled water; final concentration 20 mM) for kidney or 1 ml glycolic acid (152 mg/ml distilled water; final concentration 20 mM) for liver.
4. Incubate for 30 min at 37 °C.
5. Rinse three times with distilled water.
6. Prepare the visualization medium by dissolving 50 mg diaminobenzidine in 100 ml 0.05 M sodium acetate buffer (pH 5.3; final concentration 1.4 mM) and by adding to 100 ml medium 2 ml sodium azide (325 mg/ml distilled water; final concentration 100 mM), 1 g cobalt chloride (final concentration 42 mM) and, immediately before use, 7 µl hydrogen peroxide (stock 30 per cent; final concentration 0.6 mM).
7. Incubate for 30 min at 37 °C.
8. Rinse in distilled water.
9. Mount in glycerine–gelatin.
10. Store sections in the dark at 4 °C.
11. Control incubations should be carried out (i) in the absence of substrate; and (ii) in the presence of phenyl acetic acid, a specific inhibitor of α-hydroxy acid oxidase.

Urate oxidase (EC 1.7.3.3)

Function:	breakdown of purines
Tissues:	kidney, liver
Subcellular localization:	core of peroxisomes
pH optimum:	7.8
Validity:	qualitative (rat liver; Angermueller 1989)
Substrate:	uric acid
Other constituents:	sodium azide
Inhibitors:	2,6,8-trichloropurine, oxonic acid.

Method 27. *Cerium-diaminobenzidine-cobalt-hydrogen peroxide procedure for urate oxidase (Angermueller 1989; Frederiks et al. 1992a)*

1. Use air-dried cryostat sections.
2. Prepare 0.3 M Tris–maleate buffer (pH 7.8).
3. Prepare the incubation medium by heating the buffer to 70 °C and by adding to 100 ml buffer 3 ml cerium chloride also heated to 70 °C (371 mg/ml distilled water; final concentration 30 mM). Cool down to 37 °C and add 2 ml sodium azide (325 mg/ml distilled water; final concentration 100 mM) and 1 ml uric acid (1.9 mg/ml distilled water; final concentration 0.1 mM).
4. Incubate for 30 min at 37 °C.
5. Rinse three times with distilled water.
6. Prepare the visualization medium by dissolving 50 mg diaminobenzidine in 100 ml 0.05 M sodium acetate buffer (pH 5.3; final concentration 1.4 mM)

and by adding to 100 ml medium 2 ml sodium azide (325 mg/ml distilled water; final concentration 100 mM), 1 g cobalt chloride (final concentration 42 mM) and, immediately before use, 7 µl hydrogen peroxide (stock 30 per cent; final concentration 0.6 mM).
7. Incubate for 30 min at 37 °C.
8. Rinse in distilled water.
9. Mount in glycerine–gelatin.
10. Store sections in the dark at 4 °C.
11. Control incubations should be carried out (i) in the absence of substrate; and (ii) in the presence of 20 mM trichloropurine or oxonic acid, competitive inhibitors of urate oxidase.

4.4.3 Diazonium salt methods

In diazonium salt methods, artificial substrates are used which contain a 1-naphthol, 2-naphthol, naphthol anilic acid (naphthol AS) or naphtholamine group. These methods are used for the demonstration of activity of phosphatases (acid phosphatase; Method 28), glycosidases (Method 29), esterases (Method 29) and proteases (Methods 30 and 31). The procedure can be performed as a simultaneous coupling reaction or as a post-coupling reaction. Hexazotized *p*-rosanilin and Fast Blue BB are recommended coupling agents. The choice depends on the group of the substrate that is split off by the enzyme: 1-naphthol, 2-naphthol and naphthol AS yield precisely localized red final reaction products when coupled with hexazotized *p*-rosanilin, whereas naphtholamine should be coupled with Fast Blue BB.

Acid phosphatase (EC 3.1.3.2)

Function:	obscure
Tissues:	spleen, kidney, liver, intestine, adrenals
Subcellular localization:	lysosomes, endoplasmic reticulum
pH optimum:	5.0
Validity:	quantitative (rat liver; Frederiks *et al.* 1987c)
Substrate:	naphthol AS BI phosphoric acid
Other constituents:	–
Inhibitors:	sodium fluoride, sodium molybdate, L-tartrate.

Method 28. *Simultaneous azo-coupling method for acid phosphatase (Frederiks et al. 1987c; Van Noorden et al. 1989a)*

1. Use unfixed cryostat sections.
2. Prepare hexazotized *p*-rosanilin by dissolving 800 mg alkaline fuchsin in 2 ml 2 N HCl and 8 ml distilled water. Mix thoroughly until dissolved (2–3 h) and filter off. The solution is stable for several months when kept at 4 °C in the dark. Filter off before use. As shortly as possible before incubation, mix 1 ml

of the solution with 1 ml of an aqueous solution of sodium nitrite (80 mg/ml distilled water).
3. Prepare the incubation medium by dissolving 18 g polyvinyl alcohol in 100 ml 0.1 M sodium acetate buffer (pH 5.0) according to Method 14. To 1 ml polyvinyl alcohol-containing buffer, add 2.5 mg naphthol AS BI phosphoric acid dissolved in 25 µl dimethylformamide (final concentration 5 mM) and 80 µl hexazotized *p*-rosanilin (final concentration 10 mM).
4. Incubate for 10–60 min at 37 °C.
5. Rinse with distilled water at 60 °C.
6. Rinse with distilled water.
7. Mount in glycerine–gelatin.
8. Control incubations should be carried out (i) in the absence of substrate; and (ii) in the presence of sodium fluoride, sodium molybdate, or L-tartrate, inhibitors of acid phosphatase.

Glycosidases

Glycosidases catalyse the hydrolysis of glycoside bonds. The specificity of glycosidases is determined by the type and form of the sugar that contributes to the glycosidic bond. Glycosidases can be subdivided into polysaccharidases, oligosaccharidases and disaccharidases.

Method 29. *Simultaneous azo-coupling method for glycosidases and esterases according to Lojda et al. (1979) and Van Noorden et al. (1989a)*

1. Use unfixed crysostat sections.
2. Prepare hexazotized *p*-rosanilin by dissolving 800 mg alkaline fuchsin in 2 ml 2 N HCl and 8 ml distilled water. Mix thoroughly until dissolved (2–3 h) and filter off. The solution is stable for several months when kept at 4 °C in the dark. Filter off before use. As shortly as possible before incubation, mix 1 ml of the solution with 1 ml of an aqueous solution of sodium nitrite (80 mg/ml distilled water).
3. Prepare the incubation medium by dissolving 18 g polyvinyl alcohol in 100 ml 0.1 M sodium acetate buffer (pH 5.0) according to Method 14. Add to 1 ml polyvinyl alcohol-containing buffer, 5 mg substrate dissolved in 100 µl dimethylformamide (final concentration at least 5 mM) and 80 µl hexazotized *p*-rosanilin (final concentration 10 mM).
3. Incubate for 30–60 min at 37 °C.
4. Rinse with distilled water at 60 °C.
5. Rinse with distilled water.
6. Mount in glycerine–gelatin.

Acid α-glucosidase (EC 3.2.1.20)
Function: hydrolysis of maltose
Tissues: liver, kidney, brain, jejunum, ileum
Subcellular localization: lysosomes, brush border of enterocytes

pH optimum:	6.0
Validity:	quantitative (rat jejenum; Gutschmidt *et al.* 1979)
Substrate:	2-naphthyl-α-glucoside
Other constituents:	—
Inhibitor:	sucrose.

α-Galactosidase (EC 3.2.1.22)

Function:	degradation of complex carbohydrates
Tissues:	epithelia of epididymis, seminal vesicle, uterus, urinary bladder, trachea, bronchi, colon, thyroid gland, and nerve tissue
Subcellular localization:	lysosomes, endoplasmic reticulum
pH optimum:	5.0
Validity:	qualitative (various rat tissues; Lojda *et al.* 1979)
Substrate:	1-naphthyl-α-galactoside
Other constituents:	—
Inhibitor:	galactose.

α-Mannosidase (EC 3.2.1.24)

Function:	degradation of muco- and glycoproteins
Tissues:	epithelia of epididymis and uterus, kidney, salivary gland
Subcellular localization:	lysosomes, endoplasmic reticulum
pH optimum:	4.0–5.0
Validity:	qualitative (various rat tissues; Lojda *et al.* 1979)
Substrate:	1-naphthyl-α-mannoside
Other constituents:	—
Inhibitor:	mannonolactone.

N-Acetyl-ß-glucosaminidase (EC 3.2.1.30)

Function:	hydrolysis of oligosaccharides
Tissues:	kidney, epididymis, spleen, thyroid, liver, ovary, synovium
Subcellular localization:	lysosomes, endoplasmic reticulum, cytoplasm
pH optimum:	4.5
Validity:	qualitative (various rat tissues; Lojda *et al.* 1979) and quantitative (guinea pig thyroid; Robertson 1980)
Substrate:	naphthol AS-BI N-acetyl-β-glucosaminide
Other constituents:	—
Inhibitors:	acetate, acetamide, N-acetylglucosamine, 2-acetamide-2-deoxy-D-gluconolactone.

ß-Glucuronidase (EC 3.2.1.31)

Function:	degradation of mucopolysaccharides
Tissues:	epididymis, liver, kidney, spleen, adrenal cortex
Subcellular localization:	lysosomes, endoplasmic reticulum
pH optimum:	4.0–5.5
Validity:	quantitative (human leukocytes; Schofield *et al.* 1983)
Substrate:	naphtol AS-BI glucuronide
Other constituents:	—
Inhibitors:	D-glucuronolactone, D-saccharonolactone, citrate, ascorbic acid.

Lactase (EC 3.2.1.23)

Function:	breakdown of lactose
Tissues:	small intestine, proximal tubules of kidney
Subcellular localization:	brush border membranes
pH optimum:	5.5–6.5
Validity:	qualitative (small intestine of suckling rats; Lojda *et al.* 1979)
Substrate:	1-naphthyl-β-glucoside
Other constituents:	—
Inhibitors:	D-gluconolactone, D-galactonolactone.

Esterases

In histochemistry, the term esterases or 'non-specific' esterases is usually limited to enzymes which hydrolyse simple esters of N-free alcohols and organic acids such as carboxyl esterases (EC 3.1.1.1), aryl esterases (EC 3.1.1.2), and acetyl esterases (EC 3.1.1.6). Other esterases such as acetylcholinesterase (EC 3.1.1.7), cholinesterase (EC 3.1.1.8), lipases (EC 3.1.1.3), and phosphatases (EC 3.1.3.–) are usually dealt with separately. 'Non-specific' esterases can be demonstrated using a number of methods. These methods have been reviewed extensively by Lojda *et al.* (1979) and Hayhoe and Quaglino (1988). The latter review is limited to cytochemical methods in haematology, but esterase cytochemistry can be an important tool, particularly for diagnosis of leukaemias. The most frequently applied method is the simultaneous azo-coupling method (see Method 29) and the substrates used are α-naphthyl acetate, α-naphthyl butyrate, naphthol AS-D acetate and naphthol AS-D chloroacetate. Esterase activity is usually localized using aqueous media, but our experience is that better localization is obtained in the presence of polyvinyl alcohol. The different substrates can be split by various forms and isoenzymes of 'non-specific' esterases. Moreover, the use of different inhibitors such as eserine, diisopropyl fluorophosphate, and E600 enables discrimination between the various forms and isoenzymes. For a detailed discussion of selectivity of substrates and inhibitors we refer to Lojda *et al.* (1979) and Hayhoe and Quaglino (1988).

Non-specific esterases (EC 3.1.1.1, EC 3.1.1.2, and EC 3.1.1.6)

Function:	obscure
Tissues:	liver, kidney, intestine, blood cells
Subcellular localization:	endoplasmic reticulum, lysosomes, mitochondria
pH optimum:	5–8
Validity:	qualitative (human monocytes; Kaplow *et al.* 1976)
Substrates:	α-naphthyl acetate, α-naphthyl butyrate, naphthol AS-D acetate, naphthol AS-D chloroacetate
Other constituents:	—
Inhibitors:	physostigmine (eserine), diisopropyl fluorophosphate, diethyl-*p*-nitrophenyl phosphate (E600).

Proteases (peptidases and proteinases)

Peptidases or exopeptidases hydrolyse terminal peptide bonds, whereas proteinases or endopeptidases hydrolyse peptide bonds inside the substrate molecule. Exopeptidases and most endopeptidases can be demonstrated with a simultaneous azo-coupling method using derivatives of 1-naphthylamine, 2-naphthylamine or 4-methoxy-2-naphthylamide as substrates and Fast Blue BB (Method 30). In contrast, some endopeptidases such as cysteine proteinases are detected with a post-coupling azo method because these enzymes contain SH-groups which destroy diazonium salts rapidly (Method 31). Moreover, diazonium salts strongly inhibit the activity of proteinases non-competitively (see Van Noorden *et al.* 1989*b*).

Method 30. *Simultaneous azo-coupling method for the demonstration of peptidases according to Lojda et al. (1979) and Van Noorden et al. (1989a)*

1. Use unfixed cryostat sections.
2. Prepare the incubation medium by solving 18 g polyvinyl alcohol in 100 ml 0.1 M sodium cacodylate buffer, pH 5.5–8.0; add to 1 ml polyvinyl alcohol-containing buffer 2 mg substrate dissolved in 100 μl dimethylformamide (final concentration 3–5 mM) and 5 mg Fast Blue BB (final concentration 12.5 mM).
3. Incubate for 10 min at 37 °C.

4. Rinse with distilled water at 50 °C.
5. Chelate final reaction product with 25 mg copper sulfate in 1 ml distilled water (final concentration 100 mM) for 5 min at room temperature to insolubilize the final reaction product further and to minimize recrystallization.
6. Rinse with distilled water.
7. Mount in glycerine–gelatin.

γ-*Glutamyltransferase (EC 2.3.2.2)*

Function:	role in transport of glutathione
Tissues:	proximal tubules of kidney, capillary endothelium, exocrine pancreas, epithelium of epididymis, enterocytes, hepatocytes, epithelium of bile ducts
Subcellular localization:	plasma membranes
pH optimum:	8–9
Validity:	quantitative (various rat tissues; Gossrau 1985)
Substrate:	γ-L-glutamyl-1-naphthylamide
Other constituents:	—
Inhibitors:	*p*-chloromercuribenzoic acid, glycerol, iodoacetic acid, Mg^{2+}.

Aminopeptidase M (EC 3.4.11.2)

Function:	obscure
Tissues:	enterocytes, proximal tubules of kidney
Subcellular localization:	brush border membranes, microsomes
pH optimum:	6–9
Validity:	quantitative (pig kidney; Wachsmuth and Donner 1976)
Substrate:	leucine-2-naphthylamide
Other constituents:	—
Inhibitors:	puromycin, EDTA, 1,10-phenanthroline.

Aminopeptidase A (EC 3.4.11.7)

Function:	breakdown of angiotensin
Tissues:	glomeruli of kidney
Subcellular localization:	plasma membranes
pH optimum:	7.0
Validity:	quantitative (rat kidney; Kugler 1982*a*)
Substrate:	α-glu-4-methoxy-2-naphthylamide
Other constituents:	—
Inhibitors:	angiotensin I, II and III.

Dipeptidyl peptidase II (EC 3.4.14.2)

Function:	lysosomal catabolism of proteins
Tissues:	kidney, liver, spleen, thyroid, small intestine, vesicular gland, epididymis
Subcellular localization:	lysosomes
pH optimum:	4.5–5.5
Validity:	quantitative (rat kidney; Gossrau and Lojda 1980)
Substrate:	lys-ala-4-methoxy-2-naphthylamide
Other constituents:	–
Inhibitors:	puromycin, Tris, cations.

Dipeptidyl peptidase IV (EC 3.4.14.5)

Function:	processing of proteins, e.g. conversion of pro-interleukin-2 into interleukin-2
Tissues:	kidney, liver, intestine, lung, epididymis, seminal vesicle, salivary glands, spleen, T-helper cells
Subcellular localization:	plasma membranes
pH optimum:	7.4–8.0
Validity:	quantitative (various rat tissues; Gossrau 1985)
Substrate:	gly-pro-4-methoxy-2-naphthylamide
Other constituents:	–
Inhibitors:	diisopropyl fluorophosphate, diethyl-p-nitrophenyl phosphate (E 600).

Cathepsin B (EC 3.4.22.1)

Function:	lysosomal digestion of proteins such as collagen; activation of (pre)prohormones
Tissues:	chondrocytes, fibroblasts, synoviocytes, liver, thyroid
Subcellular localization:	lysosomes
pH optimum:	6.0
Validity:	quantitative (rat knee joints; Van Noorden et al. 1989b)
Substrate:	Z-arg-arg-4-methoxy-2-naphthylamide
Other constituents:	EDTA, dithiothreitol, L-cysteine.
Inhibitors:	leupeptin, E-64, Z-phe-ala-fluoromethylketone.

Method 31. *Post-coupling diazonium salt method for the demonstration of cathepsin B (Van Noorden et al. 1989b)*

---

1. Use unfixed and undecalcified cryostat sections of knee joints attached to transparent tape (see Method 52).

2. Dissolve 10 g polyvinyl alcohol in 100 ml 0.1 M phosphate buffer, pH 6.0 according to Method 14. Add to 1 ml polyvinyl alcohol-containing buffer 20 µl Z-arg-arg-4-methoxy-2-naphthylamide (1 mg dissolved in 10 µl dimethylformamide; final concentration 2 mM), 10 µl EDTA (38 mg/ml distilled water; final concentration 1.3 mM), 10 µl dithiothreitol (152 mg/ml distilled water; final concentration 10 mM) and 10 µl L-cysteine (33 mg/ml distilled water; final concentration 2.7 mM).
3. Incubate for 15 min at 37 °C
4. Rinse thoroughly for 5 min at 37 °C with 125 mg N-ethylmaleimide in 100 mM phosphate buffer (pH 8.0; final concentration 10 mM) to stop the reaction and to remove all SH-groups.
5. Incubate for 30 min at room temperature with 0.25 mg Fast Blue BB dissolved in 100 mM phosphate buffer (pH 8.0; final concentration 0.5 mM).
6. Rinse thoroughly in distilled water.
7. Mount in glycerine–gelatin.
8. Control incubations should be carried out (i) in the absence of substrate; and (ii) in the presence of inhibitors.

4.4.4 Indigogenic methods

In indigogenic methods products of the enzyme reaction (indoxyl or indolylamine) are oxidized to blue indigo by potassium ferricyanide. These methods are not used very frequently for the demonstration of enzyme activities because in most cases more reliable diazonium salt methods are available. Moreover, potassium ferricyanide can inhibit the activity of enzymes, e.g. ß-galactosidase in cultured fibroblasts. However, we will describe the method for the demonstration of ß-galactosidase activity in cultured cells which was validated carefully (Lund-Hansen et al. 1984).

ß-Galactosidase (EC 3.2.1.23)

Function:	degradation of mucosubstances and glycolipids
Tissues:	small intestine, fibroblasts, liver, lung, kidney, spleen, uterus, epididymis
Subcellular localization:	lysosomes
pH optimum:	3–5
Validity:	quantitative (cultured fibroblasts; Lund-Hansen et al. 1984)
Substrate:	5-bromo-4-chloro-3-indolyl-ß-D-galactopyranoside
Other constituents:	ferrocyanide, ferricyanide, NaCl
Inhibitors:	D-galactono-1,4-lactone, p-chloromercuribenzoic acid, $HgCl_2$.

Method 32. *Indigogenic method for the demonstration of β-galactosidase (Lund-Hansen et al. 1984)*

1. Use unfixed cultured cells or cryostat sections.
2. Dry cells for 20 min at 37 °C or sections for 5 min at 37 °C.
3. Prepare incubation medium by dissolving 18 g polyvinyl alcohol in 100 ml 0.1 M sodium acetate buffer (pH 4.0) according to Method 14. Add to 1 ml polyvinyl alcohol-containing buffer 10 μl potassium ferrocyanide (115 mg/ml distilled water; final concentration 3.12 mM), 10 μl potassium ferricyanide (103 mg/ml distilled water; final concentration 3.12 mM), 10 μl NaCl (580 mg/ml distilled water; final concentration 100 mM) and 20 μl 5-bromo-4-chloro-3-indolyl-β-D-galactopyranoside (150 g substrate dissolved in 20 μl 2-ethoxyethanol; final concentration 3.67 mM).
4. Incubate for 15–120 min at 37 °C.
5. Rinse with distilled water at 60 °C.
6. Rinse with distilled water.
7. Mount in glycerine–gelatin.
8. Control incubations should be carried out (i) in the absence of substrate; and (ii) in the presence of inhibitors.

4.4.5 Tetrazolium salt methods

Tetrazolium salt methods are used for the demonstration of dehydrogenases (Method 33), reductases (Methods 35 and 36), oxidases (Methods 37 and 38), and, in multistep reactions, all other classes of enzymes that can be linked to a tetrazolium salt method in the presence of auxiliary enzymes (Method 34).

Dehydrogenases

This group of enzymes has been defined as enzymes which are able to oxidize substrates by picking up electrons from the substrate. These electrons can be transferred to a coenzyme such as NAD^+ or $NADP^+$ or by other electron acceptors. All dehydrogenases can in principle be demonstrated according to one principle, with variation of their substrates and/or coenzymes. For this reason, therefore, a general procedure is described for the demonstration of dehydrogenase activities (see Method 33).

Method 33. *Tetrazolium salt method for the demonstration of dehydrogenases (Van Noorden et al. 1989a; Stoward and Van Noorden 1991)*

1. Use unfixed cryostat sections or cells.
2. Dissolve 18 g polyvinyl alcohol in 100 mM phosphate buffer (pH 7–8) or 0.1 M Tris–maleate buffer (pH 7–8)[a] according to Method 14.

3. Add to 1 ml polyvinyl alcohol-containing buffer, 10 μl of 100 × concentrated stock solutions of substrate and coenzyme (when necessary). Then add 10 μl sodium azide (33 mg/ml distilled water; final concentration 5 mM), 10 μl 1-methoxyphenazine methosulfate (15 mg/ml distilled water; final concentration 0.45 mM) or 10 μl phenazine methosulfate[b] (8 mg/ml distilled water; final concentration 0.2 mM) and 40 μl tetranitro TB or nitro BT[c] (5 mg dissolved in 20 μl ethanol and 20 μl dimethylformamide by gentle heating; final concentration 5 mM).
4. Incubate for 5–30 min at 37 °C.
5. Rinse thoroughly with 0.1 M phosphate buffer (pH 5.3) at 60 °C.
6. Mount in glycerine–gelatin.
7. Control incubations should be carried out (i) in the absence of substrate or in the absence of substrate and coenzyme; and (ii) in the presence of inhibitor.

[a] The buffer of choice is a phosphate buffer. If, however, phosphate, ADP, or ATP is involved in the reaction, the phosphate buffer must be replaced by Tris–maleate or HEPES buffer.
[b] The choice of intermediate electron carrier is determined by the localization of the dehydrogenase under study. Mitochondrial enzymes have to be demonstrated with phenazine methosulfate because 1-methoxyphenazine methosulfate cannot pass the mitochondrial membrane (Kugler 1982b). All other dehydrogenases should be detected using incubation media containing 1-methoxyphenazine methosulfate, because this electron carrier has a higher photochemical stability than phenazine methosulfate, thus resulting in lower non-specific control reactions (Van Noorden and Tas 1982).
[c] Because of the very fine formazan granules, the ease of reduction and the lack of lipid solubility of its formazan, tetranitro BT is the tetrazolium salt of choice for dehydrogenase histochemistry (Van Noorden and Butcher 1984). Only when performing kinetic measurements of formazan production in sections during incubation (see Section 2.5) is nitro BT preferred. Incubation media containing the latter tetrazolium salt remain clear for longer periods.

Glycerol-3-phosphate dehydrogenase (NAD⁺ dependent) (EC 1.1.1.8)

Function:	role in glycerol-3-phosphate shuttle
Tissues:	kidney, liver
Subcellular localization:	cytoplasm
pH optimum:	7.0–9.3
Validity:	quantitative (human leukocytes; Stuart and Simpson 1970)
Substrate:	50–100 mM glycerol-3-phosphate
Coenzyme:	3 mM NAD$^+$
Other constituents:	10 mM magnesium chloride
Inhibitors:	dihydroxyacetone phosphate, fructose-1,6-diphosphate.

Note: Kugler (1991) demonstrated that the NAD$^+$-dependent glycerol-3-phosphate dehydrogenase activity cannot be detected properly in brain sections, because the NAD$^+$-independent dehydrogenase is also demonstrated.

UDP-glucose dehydrogenase (EC 1.1.1.22)
Function:	controls rate-limiting step of the pathway for the formation of precursors involved in secondary wall synthesis in higher plants; in general synthesis of mucopolysaccharides
Tissues:	plants, liver, kidney
Subcellular localization:	cytoplasm
pH optimum:	7.5–8.5
Validity:	quantitative (plant roots; McGarry and Gahan 1985)
Substrate:	2.3 mM UDP–glucose
Coenzyme:	2.9 mM NAD^+
Other constituents:	–
Inhibitors:	UDP-D-xylulose, UDP-D-glucuronic acid.

Lactate dehydrogenase (EC 1.1.1.27)
Function:	indicator of anaerobic glycolysis
Tissues:	liver, heart, skeletal muscle, kidney
Subcellular localization:	cytoplasm, mitochondria
pH optimum:	7.5
Validity:	quantitative (rat liver, heart muscle, tracheal epithelium; Van Noorden and Vogels 1989*b*)
Substrate:	120–150 mM sodium lactate
Coenzyme:	3 mM NAD^+
Other constituents:	–
Inhibitors:	pyruvate, ATP, *p*-chloromercuribenzoic acid.

3-Hydroxybutyrate dehydrogenase (EC 1.1.1.30)
Function:	indicator of β-oxidation of fatty acids
Tissues:	liver, heart, kidney
Subcellular localization:	mitochondria
pH optimum:	8.0
Validity:	qualitative (rat liver; Rieder 1981)
Substrate:	100 mM D-3-hydroxybutyrate
Coenzyme:	3 mM NAD^+
Other constituents:	5 mM $MgCl_2$
Inhibitor:	D-lactate.

3-Hydroxyacyl CoA dehydrogenase (EC 1.1.1.35)
Function:	key enzyme in oxidation of fatty acids
Tissues:	heart, liver, skeletal muscle
Subcellular localization:	mitochondria
pH optimum:	8.5–10.0
Validity:	quantitative (rat heart; Chambers *et al.* 1982)

Substrate:	100 mM β-hydroxybutyryl-N-acetyl cysteamine or 20 mM β-hydroxybutyryl-coenzyme A
Coenzyme:	1 mM NAD$^+$
Other constituents:	4 mM nitroprusside (as capturing agent for cystamine liberated by hydrolysis of the substrate)
Inhibitor:	not known.

Malate dehydrogenase (NAD$^+$-dependent) (EC 1.1.1.37)

Function:	role in Krebs cycle; reoxidation of malate to oxaloacetate for gluconeogenesis
Tissues:	kidney, liver, heart
Subcellular localization:	cytoplasm, mitochondria
pH optimum:	7.6
Validity:	quantitative (rat liver; Wimmer and Pette 1979)
Substrate:	300 mM L-malate
Coenzyme:	3 mM NAD$^+$
Other constituents:	5 mM EDTA
Inhibitor:	(di)fluorooxalic acid.

Malate dehydrogenase or malic enzyme (NADP$^+$-dependent) (EC 1.1.1.40)

Function:	generation of NADPH for fatty acid synthesis and production of pyruvate from malate
Tissues:	liver
Subcellular localization:	cytoplasm
pH optimum:	7.2–7.5
Validity:	qualitative (rat liver; Rieder et al. 1978)
Substrate:	100 mM L-malate
Coenzyme:	0.8 mM NADP$^+$
Other constituents:	4 mM magnesium chloride
Inhibitors:	L-aspartic acid, succinic acid.

Isocitrate dehydrogenase (NAD$^+$-dependent) (EC 1.1.1.41)

Function:	role in Krebs cycle
Tissues:	kidney, liver
Subcellular localization:	mitochondria
pH optimum:	7.6
Validity:	quantitative (rat hippocampus; Kugler and Vogel 1991)
Substrate:	100 mM D,L-isocitrate
Coenzyme:	7 mM NAD$^+$
Other constituents:	10 mM magnesium chloride
Inhibitors:	3-mercapto-2-oxoglutaric acid, oxaloacetic acid.

Isocitrate dehydrogenase (NADP⁺-dependent) (EC 1.1.1.42)
Function:	generation of NADPH
Tissues:	liver, kidney
Subcellular localization:	cytoplasm
pH optimum:	7.5
Validity:	quantitative (rat hippocampus; Kugler and Vogel 1991)
Substrate:	100 mM D,L-isocitrate
Coenzyme:	4 mM NADP⁺
Other constituents:	10 mM magnesium chloride
Inhibitors:	3-mercapto-2-oxoglutaric acid, oxaloglycine.

Phosphogluconate dehydrogenase (EC 1.1.1.44)
Function:	role in hexose monophosphate shunt
Tissues:	liver, steroid hormone-producing glands, epithelium of lactating mammary gland
Subcellular localization:	cytoplasm
pH optimum:	7.5–9.0
Validity:	quantitative (rat liver; Jonges and Van Noorden 1989)
Substrate:	8 mM 6-phosphogluconic acid
Coenzyme:	0.8 mM NADP⁺
Other constituents:	5 mM magnesium chloride
Inhibitors:	ATP, p-chloromercuribenzoic acid, Hg^{2+}, iodoacetamide.

Glucose-6-phosphate dehydrogenase (EC 1.1.1.49)
Function:	key enzyme in hexose monophosphate shunt
Tissues:	liver, steroid hormone-producing glands, epithelium of lactating mammary gland
Subcellular localization:	cytoplasm
pH optimum:	7.45
Validity:	quantitative (rat liver; Van Noorden 1984)
Substrate:	10 mM glucose-6-phosphate
Coenzyme:	0.8 mM NADP⁺
Other constituents:	4 mM magnesium chloride
Inhibitors:	inorganic phosphate, glucosamine-6-phosphate.

20α-Hydroxysteroid dehydrogenase (EC 1.1.1.62)
Function:	role in synthesis of steroid hormones
Tissues:	corpus luteum, membrana granulosa of mature follicles in ovarium
Subcellular localization:	cytoplasm
pH optimum:	8.0

Validity:	quantitative (rat ovarium; Robertson *et al.* 1982)
Substrate:	1.5 mM 20α-hydroxy-4-pregnen-3-one
Coenzyme:	0.5 mM NADP⁺
Other constituents:	–
Inhibitors:	progesterone, diethylstilbestrol.

3ß-Hydroxy-Δ⁵-steroid dehydrogenase (EC 1.1.1.145)

Function:	role in synthesis of steroid hormones
Tissues:	corpus luteum
Subcellular localization:	endoplasmic reticulum and/or mitochondria
pH optimum:	9.0
Validity:	quantitative (rat ovarium; Robertson 1979; Gordon and Robertson 1986)
Substrate:	2 mM dehydroxyepiandrosterone (dissolved in dimethylformamide)
Coenzyme:	2 mM NAD⁺ or 2 mM NADP⁺
Other constituents:	3 mM EDTA
Inhibitors:	*N*-ethylmaleimide, WIN 24540.

Glycerol-3-phosphate dehydrogenase (coenzyme-independent) (EC 1.1.99.5)

Function:	role in glycerol-3-phosphate shuttle
Tissues:	small intestine, kidney, liver, heart, skeletal muscle
Subcellular localization:	mitochondria
pH optimum:	7.4
Validity:	quantitative (rat muscle, rat brain; Martin *et al.* 1985; Kugler 1991)
Substrate:	9.3 mM glycerol-3-phosphate
Coenzyme:	–
Other constituents:	–
Inhibitors:	*N*-ethylmaleimide, dihydroxyacetone phosphate.

Aldehyde dehydrogenase (EC 1.2.1.3)

Function:	detoxification
Tissues:	hepatocellular carcinoma, tongue, oesophagus, forestomach
Subcellular localization:	cytoplasm
pH optimum:	7.0
Validity:	quantitative (rat hepatocellular carcinoma; Chieco *et al.* 1986)
Substrates:	25 mM benzaldehyde, 6 mM acetaldehyde, 1.5 mM glutaraldehyde

Coenzyme:	1.5 mM NAD$^+$
Other constituents:	–
Inhibitors:	disulfiram, pyrogallol, acetophenone, ADP.

Benzaldehyde dehydrogenase (EC 1.2.1.7)
Function:	detoxification
Tissues:	hepatocellular carcinoma, tongue, oesophagus, forestomach
Subcellular localization:	cytoplasm
pH optimum:	7.0
Validity:	quantitative (rat hepatocellular carcinoma; Chieco et al. 1986)
Substrates:	25 mM benzaldehyde, 6 mM acetaldehyde, 1.5 mM glutaraldehyde
Coenzyme:	1.5 mM NADP$^+$
Other constituents:	–
Inhibitor:	disulfiram.

Glyceraldehyde-3-phosphate dehydrogenase (EC 1.2.1.12)
Function:	role in glycolytic pathway
Tissues:	liver, kidney, muscle
Subcellular localization:	cytoplasm
pH optimum:	8.5
Validity:	quantitative (mouse preovulatory oocytes; De Schepper et al. 1985)
Substrate:	2.3 mM diethyl-acetal salt of glyceraldehyde-3-phosphate
Coenzyme:	3 mM NAD$^+$
Other constituents:	–
Inhibitors:	iodoacetate, Hg^{2+}, 1,3-diphosphoglycerate.

Succinate-semialdehyde dehydrogenase (EC 1.2.1.24)
Function:	role in synthesis of neurotransmitter, 4-aminobutyric acid
Tissues:	cerebellum
Subcellular localization:	mitochondria
pH optimum:	8.4
Validity:	quantitative (rat cerebellum; Ritter 1973)
Substrates:	50 mM 4-aminobutyric acid and 30 mM α-ketoglutaric acid
Coenzyme:	1 mM NAD$^+$
Other constituents:	3 mM malonate to inhibit succinate dehydrogenase, 5 mM MgCl$_2$
Inhibitors:	p-hydroxybenzaldehyde, phenobarbital.

Succinate dehydrogenase (EC 1.3.99.1)
Function: role in Krebs cycle
Tissues: liver, heart, skeletal muscle, kidney, brain
Subcellular localization: mitochondria (inner membrane)
pH optimum: 8.0
Validity: quantitative (rat liver; Butcher 1970; Van Noorden and Vogels 1989b)
Substrate: 50 mM succinate
Coenzyme: –
Other constituents: –
Inhibitor: malonate.

Glutamate dehydrogenase (NAD$^+$-dependent) (EC 1.4.1.2)
Function: regulation of intracellular levels of ammonia
Tissues: liver, kidney
Subcellular localization: mitochondrial matrix
pH optimum: 8.0
Validity: quantitative (rat liver; Wimmer and Pette 1979; Kugler 1990a)
Substrate: 75 mM L-glutamate
Coenzyme: 3 mM NAD$^+$
Other constituents: 2 mM ADP
Inhibitors: guanosine triphosphate, adenosine triphosphate, EDTA.

Multistep reactions for enzymes other than dehydrogenases

Enzymes other than dehydrogenases can be demonstrated using tetrazolium salt methods by applying one or more auxiliary enzymes in excess. The methods are based on a chain reaction in which the product of one enzyme is used as substrate in the next reaction. The last enzyme in the chain should be a dehydrogenase which can reduce a tetrazolium salt.

The amount of auxiliary enzyme(s) needed to obtain optimum activity of the enzyme under study is dependent on many factors. Therefore, concentrations of the auxiliary enzymes have to be varied to obtain a homogeneous maximum staining of the section, and to ensure that the activity of the auxiliary enzyme(s) is not rate-limiting. The type of buffer that is recommended depends on the enzyme under study. In most cases 100 mM phosphate buffer (pH 7–8) is used, but when inorganic phosphate or ATP is involved in the enzyme reaction, it is suggested that 100 mM Tris–maleate buffer is used. As in the case of tetrazolium salt methods for dehydrogenases, a general protocol is given for enzymes demonstrated using auxiliary enzyme(s) (Method 34). The principle of the method allows quantitative determination of enzyme activities with at least one limitation. It has been shown that linear relationships can be obtained between incubation time and amounts of final reaction product (e.g. for creatine kinase, Frederiks *et al.* 1988*b*), but such a linear relationship does not exist between section thickness and formation of final reaction product. In fact, formation of final reaction product is independent of section thickness. It has been assumed that this is caused by the limited penetration of auxiliary enzyme(s) into sections. It has to be realized, therefore, that the distance of penetration into sections may differ in various tissues and in the same tissue under different (patho)-physiological conditions.

Method 34. *Tetrazolium salt method for the demonstration of enzymes in the presence of auxiliary enzyme(s) (Frederiks et al. 1987a)*

1. Dissolve lyophilized pure auxiliary enzyme(s) in distilled water and add albumin (final concentration 0.1 per cent w/v).
2. Spread 10 µl of the enzyme solution with a brush to form a film on glass slides over an area of approximately 1 cm².
3. Dry the enzyme films in air and, when dry, use slides immediately.
4. Pick up cryostat sections on the enzyme films on the glass slides.
5. Dry cryostat sections in air.
6. Prepare the incubation medium by dissolving 18 g polyvinyl alcohol in 100 ml phosphate buffer (pH 7–8) or 100 mM Tris–maleate buffer (pH 7–8) according to Method 14. Add to 1 ml 10 µl of 100 × concentrated stock solutions of substrate, co-enzyme, other constituents, and sodium azide (33 mg/ml distilled water; final concentration 5 mM), 1-methoxyphenazine methosulfate (15 mg/ml distilled water; final concentration 0.45 mM), or phenazine methosulfate (8 mg/ml distilled water; final concentration 0.2 mM) and 40 µl tetranitro BT or nitro BT (5 mg dissolved in 20 µl ethanol and 20 µl dimethylformamide by gentle heating; final concentration 5 mM).
7. Incubate sections for 5–30 min at 37 °C.
8. Rinse thoroughly with 0.1 M phosphate buffer (pH 5.3) at 60 °C.
9. Mount in glycerine–gelatin.
10. Control incubations should be carried out (i) in the absence of substrate(s); and (ii) in the presence of inhibitor(s).

Purine nucleoside phosphorylase (EC 2.4.2.1)

Function: breakdown of inosine
Tissues: endothelial cells, glial cells, fibroblasts, liver
Subcellular localization: cytoplasm
pH optimum: 8.0
Validity: quantitative (rat liver; Frederiks *et al.* 1992*b*)
Substrate: 0.5 mM inosine
Coenzyme: –
Other constituents: –

Auxiliary enzyme:	xanthine oxidase
Inhibitors:	p-chloromercuribenzoic acid, N-ethylmaleimide.

γ-Aminobutyric acid transaminase (EC 2.6.1.19)

Function:	breakdown of the neurotransmitter γ-aminobutyric acid
Tissues:	brain
Subcellular localization:	mitochondria
pH optimum:	8.0
Validity:	quantitative (rat brain; Kugler and Baier 1990)
Substrates:	50 mM γ-aminobutyric acid and 5 mM α-ketoglutaric acid
Coenzyme:	7 mM NAD^+
Other constituents:	20 mM malonate (inhibitor of succinate dehydrogenase)
Auxiliary enxymes:	succinic semialdehyde dehydrogenase, NADH dehydrogenase
Inhibitor:	3-amino-2,3-dihydrobenzoic acid.

Hexokinase (EC 2.7.1.1)

Function:	role in glycolysis
Tissues:	liver, smooth muscle cells, nerve cells, kidney, skeletal muscle, epithelial cells
Subcellular localization:	cytoplasm
pH optimum:	7.5
Validity:	quantitative (rat hippocampus; Kugler 1990b)
Substrates:	5 mM glucose, 7.5 mM ATP
Coenzyme:	1.5 mM NAD^+
Other constituents:	10 mM magnesium chloride
Auxiliary enzyme:	glucose-6-phosphate dehydrogenase
Inhibitors:	N-acetylglucosamine, lauric acid.

Note: Lawrence et al. (1989) have shown that for quantitative studies hexokinase activity can only be measured properly when an NAD^+-using glucose-6-phosphate dehydrogenase (from *Leuconostoc mesenteroides*) is applied as auxiliary enzyme and when NAD^+ is used instead of $NADP^+$ as coenzyme. When a mammalian $NADP^+$-using enzyme is used, endogenous 6-phosphogluconate dehydrogenase may also produce final reaction product and thus result in values which are too high.

Phosphofructokinase (EC 2.7.1.11)

Function:	key enzyme of glycolysis
Tissues:	liver, heart, skeletal muscle
Subcellular localization:	cytoplasm
pH optimum:	8.4
Validity:	quantitative (rat liver; Frederiks et al. 1991)
Substrates:	20 mM fructose-6-phosphate, 2 mM ATP

Coenzyme: 5.9 mM NAD⁺
Other constituents: 2 mM magnesium chloride
Auxiliary enzymes: fructose biphosphate aldolase, triose phosphate isomerase, glyceraldehyde-3-phosphate dehydrogenase
Inhibitors: 25 mM phosphoenolpyruvate, arginine phosphate, Ca^{2+}.

NAD⁺ kinase (EC 2.7.1.23)

Function: NADP⁺ formation from NAD⁺
Tissues: liver, thyroid
Subcellular localization: cytoplasm
pH optimum: 8.2
Validity: quantitative (guinea pig thyroid; Macha et al. 1975)
Substrates: 1.5 mM NAD⁺ and 2.5 mM ATP
Other constituents: 5 mM magnesium chloride, 10 mM glucose-6-phosphate
Auxiliary enzymes: glucose-6-phosphate dehydrogenase
Inhibitors: iodoacetamide, iodoacetate, adenosine diphosphate ribose, ADP, AMP.

Note: See hexokinase, p. 81.

Pyruvate kinase (EC 2.7.1.40)

Function: key enzyme of glycolysis
Tissues: liver
Subcellular localization: cytoplasm
pH optimum: 7.4
Validity: quantitative (rat liver; Klimek et al. 1988)
Substrates: 25 mM phosphoenolpyruvate, 2.5 mM ADP
Coenzyme: 1.9 mM NADP⁺
Other constituents: 25 mM magnesium chloride, 62.5 mM AMP, 25 mM glucose
Auxiliary enzymes: hexokinase, glucose-6-phosphate dehydrogenase
Inhibitors: glutathione, pyruvic acid, quercetin.

Note: See hexokinase, p. 81.

Creatine kinase (EC 2.7.3.2)

Function: catalyses synthesis and breakdown of creatine phosphate as energy buffer
Tissues: heart, skeletal muscle, brain
Subcellular localization: cytoplasm, mitochondria
pH optimum: 7.4

Validity:	quantitative (rat heart; Frederiks *et al.* 1988*b*)
Substrates:	20 mM phosphocreatine and 1.0 mM ADP
Coenzyme:	0.45 mM NADP$^+$
Other constituents:	4 mM AMP, 10 mM glucose, 10 mM magnesium chloride
Auxiliary enzymes:	hexokinase, glucose-6-phosphate dehydrogenase
Inhibitors:	dinitrofluorobenzene, acetate, adenosine.

Note: See hexokinase, p. 81.

Guanine deaminase (EC 3.5.4.3)

Function:	hydrolytic deamination of guanine to xanthine
Tissues:	liver, brain, kidney
Subcellular localization:	cytoplasm
pH optimum:	7.8
Validity:	qualitative (human liver; Ito *et al.* 1988)
Substrate:	0.5 mM guanine
Coenzyme:	–
Other constituents:	0.1 M *N*-bis(2-hydroxyethyl)glycine
Auxiliary enzyme:	xanthine oxidase
Inhibitor:	5-aminoimidazole-4-carboxamide.

Fructose-biphosphate aldolase (EC 4.1.2.13)

Function:	role in glycolysis
Tissues:	heart, skeletal muscle, liver, erythrocytes
Subcellular localization:	cytoplasm
pH optimum:	7–8
Validity:	qualitative (rat liver; Lojda *et al.* 1979)
Substrate:	25 mM D-fructose-1,6-diphosphate
Coenzyme:	1.5 mM NAD$^+$
Other constituents:	–
Auxiliary enzymes:	triose phosphate isomerase, glyceraldehyde-3-phosphate dehydrogenase
Inhibitors:	ADP, EDTA, fructose-1-phosphate, *O*-phenanthroline.

Glucose-6-phosphate isomerase (EC 5.3.1.9)

Function:	role in glycolysis
Tissues:	heart, skeletal muscle, liver, kidney, spleen, pancreas, oocytes
Subcellular localization:	cytoplasm
pH optimum:	7.4
Validity:	quantitative (mouse preovulatory oocytes; De Schepper *et al.* 1985)

Substrate: 0.7 mM fructose-6-phosphate
Coenzyme: 0.5 mM NADP$^+$
Other constituents: 4 mM magnesium chloride
Auxiliary enzyme: glucose-6-phosphate dehydrogenase
Inhibitor(s): 6-phosphogluconate, phosphoenolpyruvic acid.

Note: See hexokinase, p. 81.

Phosphoglucomutase (EC 5.4.2.2)

Function: glycogen metabolism
Tissues: heart, skeletal muscle, liver, kidney, spleen, pancreas
Subcellular localization: cytoplasm
pH optimum: 7.4
Validity: qualitative (rat heart; De Vries and Meijer 1976)
Substrate: 0.6 mM glucose-1-phosphate
Coenzyme: 0.7 mM NADP$^+$
Other constituents: 2 mM magnesium chloride
Auxiliary enzyme: glucose-6-phosphate dehydrogenase
Inhibitor: glucose-1,6-diphosphate.

Reductases

In the past, reductase activity was indiscriminately demonstrated in histochemistry as diaphorase activity. Diaphorases are defined as enzymes capable of catalysing the oxidation of either NADH or NADPH in the presence of artificial electron acceptors such as dyes, ferricyanide, and quinones. Since many flavoproteins can be reduced by NAD(P)H, and since practically all reduced flavoproteins are capable of reducing artificial acceptors in turn, a number of histochemical methods have been described for visualizing the activity of such 'diaphorases'. However, in the past decade, it has become clear that particular reductases could be demonstrated histochemically with the use of specific reaction conditions and inhibitors. Histochemical procedures have been described for two of these reductases based on the use of a tetrazolium salt as electron acceptor and NADPH or NADH as substrate. Techniques for the demonstration of these reductases are described in Methods 35 and 36.

NADPH-ferrohaemoprotein reductase (formerly called NADPH: cytochrome c (P450) reductase) (EC 1.6.2.4)

Function: one-electron transfer to cytochrome P450
Tissues: liver
Subcellular localization: endoplasmic reticulum

Validity:	quantitative (rat liver; Van Noorden and Butcher 1986)
Substrate:	NADPH
Other constituents:	–
Inhibitors:	NADP⁺, *p*-chloromercuribenzoic acid.

Method 35. *Tetrazolium salt method for the demonstration of NADPH-ferrohaemoprotein reductase (NADPH:cytochrome c (P450) reductase; Van Noorden and Butcher 1986)*

1. Use unfixed cryostat sections.
2. Dissolve 18 g polyvinyl alcohol in 100 ml 0.1 M phosphate buffer (pH 7.4) according to Method 14.
3. To 1 ml polyvinyl alcohol-containing medium, add 10 µl NADPH (42 mg/ml distilled water; final concentration 0.5 mM) and 40 µl tetranitro BT (5 mg dissolved in 20 µl ethanol and 20 µl dimethylformamide by gentle heating; final concentration 5mM).
4. Incubate for 5–10 min at 37 °C.
5. Rinse thoroughly with 0.1 M phosphate buffer (pH 5.3) at 60 °C.
6. Mount in glycerine–gelatin.
7. Control incubations should be carried out (i) in the absence of substrate; (ii) in the presence of NADP⁺; and (iii) in the presence of dicumarol (which should have no effect).

NAD(P)H dehydrogenase or D,T-diaphorase (EC 1.6.99.2)

Function:	two-electron transfer to cytochromes
Tissues:	liver
Subcellular localization:	cytoplasm
Validity:	quantitative (fish motoneurons; Straatsburg *et al.* 1989)
Substrate:	0.5 mM NAD(P)H
Other constituents:	1.0 mM menadione
Inhibitor:	dicumarol.

Method 36. *Tetrazolium salt method for the demonstration of NAD(P)H dehydrogenase (Straatsburg et al. 1989)*

1. Use unfixed cryostat sections.
2. Dissolve 18 g polyvinyl alcohol in 100 ml 0.1 M phosphate buffer (pH 7.4) according to Method 14.
3. To 1 ml polyvinyl alcohol-containing medium, add 10 µl NADPH or NADH (42 mg/ml distilled water; final concentration 0.5 mM), 10 µl menadione (17 mg/ml distilled water; final concentration 1.0 mM), and 40 µl tetranitro BT (5 mg dissolved in 20 µl ethanol and 20 µl dimethylformamide by gentle heating; final concentration 5 mM).

4. Incubate for 5–10 min at 37 °C.
5. Rinse thoroughly with 0.1 M phosphate buffer (pH 5.3) at 60 °C.
6. Mount in glycerine–gelatin.
7. Control incubations should be carried out (i) in the absence of substrate; (ii) in the presence of dicumarol; and (iii) in the presence of NADP$^+$ (should have no effect).

Oxidases

Oxidases are characterized by their use of oxygen as substrate; the electrons produced can also be accepted by tetrazolium salts, thus producing formazan (see Methods 37 and 38).

Xanthine oxidoreductase (xanthine:NAD$^+$, EC 1.1.1.204 and xanthine: O$_2$, EC 1.2.1.37)

Function: production of uric acid (a radical scavenger?)
Tissue: liver, duodenum, epithelia
Subcellular localization: cytoplasm, peroxisomes
Validity: qualitative (rat liver; Kooij *et al.* 1991)
Substrate: 0.5 mM hypoxanthine
Other constituents: —
Inhibitor: allopurinol.

Method 37. *Tetrazolium salt method for the demonstration of xanthine oxidoreductase (Kooij et al. 1991)*

1. Use unfixed cryostat sections.
2. Dissolve 18 g polyvinyl alcohol in 0.1 M phosphate buffer (pH 8.0) according to Method 14.
3. To 1 ml polyvinyl alcohol-containing medium add 10 µl hypoxanthine (6.8 mg/ml 0.25 N NaOH, final concentration 0.5 mM), 10 µl 1-methoxyphenazine methoxysulfate (15 mg/ml distilled water; final concentration 0.45 mM) and 40 µl tetranitro BT (1 mg tetranitro BT dissolved in 20 µl ethanol and 20 µl dimethylformamide by gentle heating; final concentration 1 mM).
4. Incubate fresh sections immediately after cutting for 30 min at 37 °C.
5. Rinse thoroughly with 0.1 M phosphate buffer (pH 5.3) at 60 °C.
6. Mount in glycerine–gelatin.
7. Control incubations should be carried out (i) in the absence of substrate; and (ii) in the presence of allopurinol.

Monoamine oxidase (EC 1.4.3.4)

Function: 'inactivation of neurotransmitters
Tissues: liver, brain

Subcellular localization: outer mitochondrial membrane
Validity: quantitative (rat liver; Frederiks and Marx 1985)
Substrate: tryptamine
Other constituents: —
Inhibitors: iproniazid phosphate, pargylin.

Method 38. *Tetrazolium salt method for the demonstration of monoamine oxidase (Frederiks and Marx 1985)*

1. Use unfixed cryostat sections.
2. Prepare the incubation medium by adding to 1 ml 25 mM phosphate buffer (pH 8.0), 10 µl tryptamine–HCl (125 mg dissolved in 600 µl dimethylformamide and 400 µl ethanol; final concentration 6.25 mM), 10 µl phenazine methosulfate (8 mg/ml distilled water; final concentration 0.2 mM) and 10 µl tetranitro BT (0.25 mg dissolved in 5 µl ethanol and 5 µl dimethylformamide by gentle heating; final concentration 0.25 mM).
3. Incubate for 30–60 min at 37 °C.
4. Rinse thoroughly with 0.1 M phosphate buffer (pH 5.3) at 60 °C.
5. Mount in glycerine–gelatin.
6. Control incubations should be carried out (i) in the absence of substrate; and (ii) in the presence of inhibitor.

4.4.6 Indoxyl-tetrazolium salt methods

Alkaline phosphatase (EC 3.1.3.1)

Function: involved in active transport processes and bone formation
Tissues: liver, small intestine, kidney, periosteum
Subcellular localization: plasma membrane
Validity: quantitative (rat liver; Van Noorden and Jonges 1987)
Substrate: 5-bromo-4-chloro-3-indolyl-phosphate
Other constituents: magnesium chloride
Inhibitor: L-4-bromotetramisole.

Method 39. *Indoxyl-tetranitro BT method for the demonstration of alkaline phosphatase (Van Noorden and Jonges 1987)*

1. Use unfixed cryostat sections.
2. Dissolve 18 g polyvinyl alcohol in 0.1 M Tris–HCl buffer (pH 9.0) according to Method 14.
3. Add to 1 ml polyvinyl alcohol-containing medium, 7 µl 5-bromo-4-chloro-3-indolyl-phosphate (1 mg in 20 µl dimethylformamide; final concentration

0.7 mM), 10 µl 1-methoxyphenazine methosulfate (15 mg/ml distilled water; final concentration 0.45 mM), 10 µl MgCl$_2$ (204 mg/ml distilled water; final concentration 10 mM), 10 µl sodium azide (34 mg/ml distilled water; final concentration 5 mM) and 40 µl tetranitro BT (5 mg dissolved in 20 µl ethanol and 20 µl dimethylformamide by gentle heating; final concentration 5 mM).
4. Incubate for 15 min at 37 °C.
5. Rinse thoroughly with 0.1 M phosphate buffer (pH 5.3) at 60 °C.
6. Mount in glycerine–gelatin.
7. Control incubations should be carried out (i) in the absence of substrate; and (ii) in the presence of tetramisole.

4.4.7 Diaminobenzidine methods

Diaminobenzidine methods are used for the demonstration of catalase, peroxidase, and cytochrome c oxidase activity (see Methods 40–42). The consequence of the use of diaminobenzidine methods is that in all cases pseudoperoxidases (haemoglobin) and myeloperoxidases will interfere. Neither of these nonspecific reactions can be prevented by high substrate concentrations, fixation, or conventional inhibitors. It has to be taken into account, therefore, that these reactions may take place in certain types of cells when demonstrating catalase, peroxidase, or cytochrome c oxidase activity.

Catalase (EC 1.11.1.6)

Function:	scavenging of hydrogen peroxide
Tissues:	liver, kidney
Subcellular localization:	(micro)peroxisomes, cytoplasm of erythrocytes
pH optimum:	10.5
Validity:	qualitative (rat liver; Angermueller and Fahimi 1981)
Substrate:	hydrogen peroxide
Other constituents:	–
Inhibitor:	3-amino-1,2,4-triazole.

Method 40. *Diaminobenzidine method for the demonstration of catalase activity (Roels et al. 1975; Angermueller and Fahimi 1981)*

1. Use cryostat sections.
2. Fix sections in 0.3 per cent (w/v) glutaraldehyde in distilled water for 5 min at room temperature.
3. Rinse three times in distilled water.
4. Prepare incubation medium by dissolving 2 g polyvinyl alcohol in 100 ml 0.1 M glycine–NaOH buffer (pH 10.5) according to Method 14. Add 76 mg

diaminobenzidine (final concentration 5 mM) and 200 µl hydrogen peroxide (stock solution 30 per cent; final concentration 0.06 per cent or 18 mM).
5. Incubate sections for 30 min at 37 °C.
6. Rinse thoroughly in distilled water.
7. Mount in glycerine–gelatin.
8. Control incubations should be carried out (i) in the absence of substrate; and (ii) by preincubating in the presence of 50 mM 3-amino-1,2,4-triazole and 18 mM hydrogen peroxide.

Peroxidase (EC 1.11.1.7)

Function:	scavenging of hydrogen peroxide
Tissues:	Kupffer cells, neutrophils, eosinophils
Subcellular localization:	cytoplasmic granules, endoplasmic reticulum
pH optimum:	6.5
Validity:	qualitative (rat liver; Angermueller and Fahimi 1981; Hayhoe and Quaglino 1988)
Substrate:	hydrogen peroxide
Other constituents:	–
Inhibitor:	high concentrations of hydrogen peroxide.

Method 41. *Diaminobenzidine method for the demonstration of peroxidase activity (Roels et al. 1975; Angermueller and Fahimi 1981)*

1. Use unfixed cryostat sections.
2. Prepare the incubation medium by dissolving 2 g polyvinyl alcohol in 100 ml 0.1 M cacodylate buffer (pH 6.5) according to Method 14. Add 38 mg diaminobenzidine (final concentration 2.5 mM) and 7 µl hydrogen peroxide (stock solution 30 per cent; final concentration 0.002 per cent or 0.7 mM).
3. Incubate for 5 min at room temperature.
4. Rinse thoroughly in distilled water.
5. Mount in glycerine–gelatin.
6. Control incubations should be carried out (i) in the absence of substrate; and (ii) by preincubating in the presence of 18 mM hydrogen peroxide.

Cytochrome c oxidase (EC 1.9.3.1)

Function:	terminal step of mitochondrial electron transfer
Tissues:	liver, heart, brain
Subcellular localization:	mitochondria
pH optimum:	7.4
Validity:	quantitative (rat brain; Angermueller and Fahimi 1981; Hiraoka *et al.* 1986; Kugler *et al.* 1988*b*)
Substrate:	cytochrome *c*
Other constituents:	–
Inhibitor:	KCN

Method 42. *Diaminobenzidine method for the demonstration of cytochrome c oxidase (Kugler et al. 1988b)*

1. Use unfixed cryostat sections.
2. Prepare incubation medium by dissolving 2 g polyvinyl alcohol in 100 ml 0.1 M phosphate buffer (pH 7.4) according to Method 14. Add 114 mg diaminobenzidine (final concentration 7.5 mM) and 50 mg cytochrome *c* (final concentration 0.4 mM).
3. Incubate for 5 min at room temperature.
4. Rinse thoroughly in distilled water.
5. Mount in glycerine–gelatin.
6. Control incubations should be carried out (i) in the absence of substrate; and (ii) in the presence of 10 mM KCN.

4.4.8 Fluorescence methods with 5'-nitrosalicylaldehyde

Elastase (EC 3.4.21.11)
Function:	breakdown of collagen
Tissues:	neutrophils
Subcellular localization:	cytoplasmic granules
Validity:	semi-quantitative (hamster neutrophils; Rudolphus *et al.* 1992)
Substrate:	Suc-ala-ala-ala-4-methoxy-2-naphthylamide
Other constituents:	0.5 M NaCl
Inhibitors:	secretory leukocyte proteinase inhibitor (SLPI), antileukoprotease (ALP), L658.758.

Cathepsin B (EC 3.4.22.1)
Function:	lysosomal digestion of proteins such as collagen; activation of (pre)prohormones
Tissues:	fibroblasts, chondrocytes, synoviocytes, bone marrow cells, liver
Subcellular localization:	lysosomes
Validity:	qualitative (human fibroblasts; Van Noorden *et al.* 1987)
Substrate:	*N*-CBZ-ala-arg-arg-4-methoxy-2-naphthylamide
Other constituents:	EDTA, Dithiothreitol, L-cysteine
Inhibitors:	leupeptin, E–64, Z-phe-ala-fluoromethylketone.

Method 43. *Fluorescence method with 5'-nitrosalicylaldehyde for the demonstration of proteases (Van Noorden et al. 1987; Rudolphus et al. 1992)*

1. Dry unfixed cryostat sections or cell preparations in air for 10 min.
2b. To demonstrate elastase activity, prepare incubation medium by adding to 1 ml 100 mM Tris–HCl buffer (pH 7.2), 30.5 mg NaCl (final concentration 0.5 M), 10 µl 2-hydroxy-5-nitrobenzaldehyde (5'-nitro-salicylaldehyde; 17 mg dissolved in 200 µl ether and 800 µl dimethylformamide; final concentration 1 mM) and 1 mg Suc-ala-ala-ala-4-methoxy-2-naphthylamide dissolved in 20 µl dimethylformamide (final concentration 2 mM).
2b. To demonstrate cathepsin B activity, prepare incubation medium by adding to 100 mM phosphate buffer (pH 6.0), 10 µl dithiothreitol (15 mg/ml distilled water; final concentration 1 mM), 10 µl EDTA (50 mg/ml distilled water; final concentration 1.3 mM) 10 µl L-cysteine (32 mg/ml distilled water; final concentration 2.7 mM), 10 µl 2-hydroxy-5-nitrobenzaldehyde (5'-nitrosalicylaldehyde; 17 mg dissolved in 200 µl ether and 800 µl dimethylformamide; final concentration 1 mM) and 1 mg *N*-CBZ-ala-arg-arg-4-methoxy-2-naphthylamide dissolved in 20 µl dimethylformamide (final concentration 2 mM).
3. Pour a drop of incubation medium on to the section or cell preparation (50–100 µl) to start the reaction and monitor incubation time.
4. Put cover slip on top of medium and attach it to the object glass using a drop of a warm solution of gelatin in distilled water (8 per cent (w/v)) at each corner of the cover slip.
5. Use a fluorescence microscope with an excitation filter of 440–500 nm and a barrier filter of 515 nm.
6. Monitor development of yellow fluorescence as a product of protease activity by incubating at room temperature on the stage of the fluorescence (photo)microscope for time intervals up to 120 min. Green fluorescence is due to non-specific binding of 5'-nitrosalicylaldehyde to proteins.
7. Take photomicrographs during incubation.
8. Control incubations should be carried out in the presence of inhibitors.

Note: Monitoring of the formation of fluorescence is necessary because the final reaction product diffuses and recrystallizes rapidly during rinsing and mounting of preparations.

4.4.9 Thiocholine methods

The original thiocholine method of Karnovsky and Roots (1964) is still used with minor modifications for the demonstration of acetylcholinesterase activity (see Method 44).

Acetylcholinesterase (EC 3.1.1.7)

Function:	breakdown of acetylcholine after neurotransmission in synapses
Tissues:	brain, nerves
Subcellular localization:	membrane-bound, cytoplasm
pH optimum:	7.4

Validity:	quantitative (rat brain; Andrä and Lojda 1986)
Substrate:	acetylthiocholine iodide
Other constituents:	sodium citrate, copper sulfate, potassium ferricyanide
Inhibitors:	physostigmine, diisopropyl fluorophosphate, tetraisopropyl phosphoramide, 1,5-*bis*-(4-trimethylammoniumphenyl)-pentane-3-on-diiodide (62C47).

Method 44. *Thiocholine method for the demonstration of acetylcholinesterase (Kugler 1987)*

1. Use unfixed cryostat sections.
2. Dissolve 18 g polyvinyl alcohol in 100 ml HEPES buffer (pH 7.0) according to Method 14.
3. To 1 ml polyvinyl alcohol-containing medium, add (strictly in the following order) 10 µl sodium citrate (882 mg/ml distilled water; final concentration 30 mM), 10 µl copper sulfate (450 mg/ml distilled water; final concentration 18 mM), 10 µl potassium ferricyanide (127 mg/ml distilled water; final concentration 3 mM), and 10 µl acetylthiocholine iodide (88 mg/ml distilled water; final concentration 3 mM).
4. Adjust the pH of the medium to pH 6.0.
5. Incubate for 60 min at 37 °C.
6. Rinse thoroughly with distilled water.
7. Mount in glycerine–gelatin.
8. Control incubations should be carried out (i) in the absence of substrate; and (ii) in the presence of inhibitor(s).

4.4.10 Synthesis reactions

An example of a synthesis reaction is the method to demonstrate glycogen phosphorylase activity (Method 45). During the first incubation step glycogen is produced from glucose-1-phosphate. In the second step glycogen is stained using the periodic acid–Schiff (PAS) reaction. The semipermeable membrane technique is used to demonstrate glycogen phosphorylase activity. The product of the first incubation step, glycogen, dissolves easily in aqueous media; this can be prevented by combining the semipermeable membrane with a gelled incubation medium.

Glycogen phosphorylase (EC 2.4.1.1)

Function:	breakdown of glycogen
Tissues:	liver, heart, skeletal muscle
Subcellular localization:	endoplasmic reticulum
pH optimum:	5.8

Validity: quantitative (rat liver; Frederiks et al. 1987b)
Substrate: glucose-1-phosphate
Other constituents: AMP, sodium fluoride, EDTA
Inhibitor: *p*-chloromercuribenzoic acid.

Method 45. *Semipermeable membrane technique for the demonstration of glycogen phosphorylase (Frederiks et al. 1987b)*

1. Prepare incubation vessels consisting of a semipermeable membrane stretched over a Perspex ring as described in Method 15.
2. Prepare incubation medium by heating to 60 °C 2 ml 3 per cent (w/v) agar in 0.1 M sodium acetate buffer (pH 5.8) and by adding a mixture of the following solutions which is also heated to 60 °C: 1.2 ml 0.1 M sodium acetate buffer (pH 5.8), 320 µl glucose-1-phosphate (280 mg/ml distilled water; final concentration 60 mM), 160 µl EDTA (60 mg/ml distilled water; final concentration 6.5 mM), 160 µl NaF (30 mg/ml distilled water; final concentration 28.6 mM), and 160 µl AMP (77 mg/ml distilled water; final concentration 6.1 mM).
3. Pour the solution into incubation vessels and let the medium solidify for 5 min at room temperature.
4. Adjust the temperature of the incubation vessels to 37 °C for 5 min.
5. Allow the sections to adhere to semipermeable membranes.
6. Incubate for 10 min at 37 °C with sections on top of the vessels.
7. Remove the gelled incubation medium with a spatula.
8. Clean membranes with tissue paper.
9. Dry sections for 60 min at room temperature.
10. Incubate for 20 min in a 0.5 per cent (w/v) periodic acid solution in distilled water.
11. Rinse three times in distilled water.
12. Incubate for 30 min in Schiff's reagent at room temperature. Prepare solution by dissolving 0.5 g basic fuchsin in 15 ml 1 N HCl and when it is completely dissolved, add 85 ml of a 0.6 per cent potassium bisulfite solution. Allow solution to stand at room temperature. Add 0.3 g Norit (activated charcoal) and stir for 5 min. Filter off and allow solution to stand for at least 24 h at 4 °C. The solution must be completely colourless. Store at 4 °C. The solution is stable for about 2 months.
13. Rinse for 20 min in tap water.
14. Rinse three times in distilled water.
15. Mount in glycerine–gelatin.

4.4.11 Natural chromophores

The product of monophenol monooxygenase or DOPA-oxidase is melanin which is directly visible in light microscopical preparations (Method 46).

Monophenol monoxygenase (EC 1.14.18.1)

Function:	synthesis of melanin
Tissues:	melanocytes of epidermis, certain ganglion cells, melanoma cells
Subcellular localization:	cytoplasm
pH optimum:	7.2
Validity:	quantitative (cultured embryos of *Ciona intestinalis*; Whittaker 1981)
Substrates:	L-dihydroxyphenylalanine, tyrosine, oxygen
Other constituents:	–
Inhibitors:	cyanide, diethyldithiocarbonate, cysteine, glutathione.

Method 46. L-*DOPA reaction for the demonstration of monophenol monoxygenase (Whittaker 1981)*

1. Fix cultured embryos or cryostat sections for 2 h in 70 per cent ethanol at 4 °C.
2. Prepare incubation medium by dissolving 79 mg L-dihydrophenylalanine (final concentration 4 mM) in 100 ml 0.1 M phosphate buffer (pH 7.2).
3. Incubate for 1–8 h at 37 °C. Renew the incubation medium every 2 h.
4. Rinse in distilled water after incubation.
5. Post-treat with 35 per cent ethanol in distilled water (v/v) for 6 h at room temperature to remove unreacted substrate.
6. Rinse in distilled water.
7. Mount in glycerine–gelatin.

Cytochrome P450 receives reducing equivalents via one-electron transfer from NADPH: cytochrome *c* (P450) reductase. It shows a specific Soret band at 450 nm in its reduced form and this absorbance at 450 nm can be used as a measure of the amount of cytochrome P450 present (see Method 47). Because of interference of haem proteins such as haemoglobin and methaemoglobin in the absorbance of P450, proper control reactions have to be subtracted from the test reaction. Moreover, differences between the absorbance at 450 nm and that at 490 nm have to be calculated for the analysis of amounts of P450 in sections or cells. It also implies that this method can only be used in a quantitative manner.

Cytochrome P450 (no EC number assigned)

Function:	metabolism of drugs, carcinogens, and steroids
Tissues:	liver, ovary, adrenal
Subcellular localization:	endoplasmic reticulum
Validity:	quantitative (rat liver; Watanabe *et al.* 1989)

Method 47. *Cytophotometric method for the measurement of cytochrome P450 (Watanabe et al. 1989)*

1. Cut series of four serial cryostat sections of 10–40 μm thickness for each measurement.
2. Incubate the sections separately for 1 min at 37 °C in the following media:
   A. 10 ml 50 mM Tris–HCl (pH 8.0) containing 1 g sucrose (final concentration 292 mM).
   B. 10 ml 50 mM Tris–HCl (pH 8.0) containing 1 g sucrose (final concentration 292 mM) after being saturated with carbon monoxide.
   C. 10 ml 50 mM Tris–HCl (pH 8.0) containing 1 g sucrose (final concentration 292 mM) and 50 mg dithionite (final concentration 24 mM).
   D. 10 ml 50 mM Tris–HCl (pH 8.0) containing 1 g sucrose (final concentration 292 mM) and 50 mg dithionite (final concentration 24 mM) saturated with carbon monoxide.
3. Replace media with fresh media and mount the sections.
4. Take absorbance readings from each section at 450 and 490 nm with a band width of less than 1.3 nm and no longer than 5 min after mounting.
5. Calculate the concentration of P450 in the tissue studied with the use of the following equations:

$$\text{concentration (moles/litre)} = \frac{(\Delta \text{MIA}_{450-490})}{\varepsilon} \times \frac{D}{d}$$

where $\Delta \text{MIA}_{450-490}$ = absorbance due to P450; this is given by $(A_{450-490}) + (B_{490-450}) + (C_{490-450}) + (D_{450-490})$ (where A–D are the absorbances of sections A–D); D and d = length of light path (D, 1 cm; d, section thickness in cm, see also Method 3); and ε = extinction coefficient (91 000 $\text{l·moles}^{-1}\text{·cm}^{-1}$).

References

Altman, F. P. and Chayen, J. (1965). Retention of nitrogenous material in unfixed sections during incubations for histochemical demonstration of enzymes. *Nature*, **207**, 1205–6.

Andrä, J. and Lojda, Z. (1986). A histochemical method for the demonstration of acetylcholinesterase activity using semipermeable membranes. *Histochemistry*, **84**, 575–9.

Angermueller, S. (1989). Peroxisomal oxidases: cytochemical localization and biological relevance. *Progress in Histochemistry and Cytochemistry*, **20**(1), 1–65.

Angermueller, S. and Fahimi, H. D. (1981). Selective cytochemical localization of peroxidase, cytochrome oxidase and catalase in rat liver with 3,3-diaminobenzidine. *Histochemistry*, **71**, 33–44.

Angermueller, S. and Fahimi, H. D. (1988). Light microscopic visualization of the reaction product of cerium used for localization of peroxisomal oxidases. *Journal of Histochemistry and Cytochemistry*, **36**, 23–8.

Angermueller, S., Leupold, C., Voelkl, A., and Fahimi, H. D. (1986). Electron

microscopic cytochemical localization of α-hydroxyacid oxidase in rat liver. Association with the crystalline core and matrix of peroxisomes. *Histochemistry*, **85**, 403–9.
Briggs, R. T., Drath, D. B., Karnovsky, M. L., and Karnovsky, M. J. (1975). Localization of NADH oxidase on the surface of human polymorphonuclear leukocytes by a new cytochemical method. *Journal of Cell Biology*, **67**, 566–86.
Butcher, R. G. (1970). Studies on succinate oxidation. I. The use of intact tissue sections. *Experimental Cell Research*, **60**, 54–60.
Chambers, D. J., Braimbridge, M. V., Frost, G. T. B., Nahir, A. M., and Chayen, J. (1982). A quantitative cytochemical method for the measurement of β-hydroxyacyl CoA dehydrogenase activity in rat heart muscle. *Histochemistry*, **75**, 67–76.
Chieco, P., Normanni, P., Moslen, M. T., and Maltoni, C. (1986). Quantitative histochemistry of benzaldehyde dehydrogenase in hepatocellular carcinomas of vinyl chloride-treated rats. *Journal of Histochemistry and Cytochemistry*, **34**, 151–8.
De Schepper, G. G., Van Noorden, C. J. F., and Koperdraad, F. (1985). A cytochemical method for measuring enzyme activity in individual preovulatory mouse oocytes. *Journal of Reproduction and Fertility*, **74**, 709–16.
De Vries, G. P. and Meyer, A. E. F. H. (1976). Semipermeable membranes for improving the histochemical demonstration of enzyme activities in tissue sections. VI. D-glucose-6-phosphate isomerase and phosphoglucomutase. *Histochemistry*, **50**, 1–8.
Dodds, R. A., Pitsillides, A. A., and Frost, G. T. B. (1990). A quantitative cytochemical method for ornithine decarboxylase activity. *Journal of Histochemistry and Cytochemistry*, **38**, 123–7.
Ernst, S. A. (1972). Transport adenosine triphosphatase cytochemistry. II. Cytochemical localization of ouabain-sensitive, potassium-dependent phosphatase activity in the secretory epithelium of the avian salt gland. *Journal of Histochemistry and Cytochemistry*, **20**, 23–38.
Firth, J. A. (1987). Quantitative assessment of NA^+, K^+-ATPase localization by direct and indirect *p*-nitrophenyl phosphatase methods. *Journal of Histochemistry and Cytochemistry*, **35**, 507–13.
Frederiks, W. M. and Marx, F. (1985). Quantitative aspects of the histochemical tetrazolium salt reaction of monoamine oxidase in rat liver. *Histochemical Journal*, **17**, 707–15.
Frederiks, W. M. and Marx, F. (1988). A quantitative histochemical study of 5′-nucleotidase activity in rat liver using the lead salt method and polyvinyl alcohol. *Histochemical Journal*, **20**, 207–14.
Frederiks, W. M., Marx, F., and Strootman, F. (1987a). Demonstration of creatine kinase in myocardial and skeletal muscle using the semipermeable membrane technique. *Journal of Molecular and Cellular Cardiology*, **19**, 311–17.
Frederiks, W. M., Marx, F., and Van Noorden, C. J. F. (1987b). Quantitative histochemical assessment of heterogeneity of glycogen phosphorylase activity in liver parenchyma from fasted rats using the semipermeable membrane technique and the PAS-reaction. *Histochemical Journal*, **19**, 150–6.
Frederiks, W. M., Marx, F., Jonges, G. N., and Van Noorden, C. J. F. (1987c). Quantitative histochemical study of acid phosphatase activity in rat liver using a semipermeable membrane technique. *Journal of Histochemistry and Cytochemistry*, **35**, 175–80.

Frederiks, W. M., Marx, F., and Van Noorden, C. J. F. (1988a). Histochemical demonstration of creatine kinase activity using polyvinyl alcohol and auxiliary enzymes. *Histochemical Journal*, **19**, 529–32.

Frederiks, W. M., Marx, F., and Van Noorden, C. J. F. (1988b). Quantitative histochemistry of creatine kinase in rat myocardium and skeletal muscle. *Histochemical Journal*, **20**, 624–8.

Frederiks, W. M., Patel, H. R. H., Marx, F., Gossrau, R., Kooij, A., and Van Noorden, C. J. F. (1990). Light microscopical detection of D-amino acid oxidase activity in unfixed cryostat sections of rat kidney and liver using the cerium-DAB-cobalt-H_2O_2 procedure and a semipermeable membrane. *Acta Histochemica*, Supplement **40**, 95–100.

Frederiks, W. M., Marx, F., and Van Noorden, C. J. F. (1991). Homogeneous distribution of phosphofructokinase in the rat liver acinus. A quantitative histochemical study. *Hepatology*, **14**, 634–9.

Frederiks, W. M., Bosch, K. S., and Van Gulik, T. (1992b). A quantitative histochemical procedure for the demonstration of purine nucleoside phosphorylase activity in rat and human liver using tetranitro BT and xanthine oxidase as auxiliary enzyme. *Histochemical Journal*, **24**, in press.

Frederiks, W. M., Van Noorden, C. J. F., Marx, F., Gallagher, P. T., and Swann, B. P. (1992a). *In situ* kinetic measurements of D-amino acid oxidase in rat liver with respect to its substrate specificity. *European Journal of Cell Biology*, in press.

Gordon, M. and Robertson, W. R. (1986). The application of continuous monitoring microdensitometry to an analysis of NAD^+ binding and 3β-hydroxysteroid dehydrogenase activity in the regressing corpus luteum of the proestrous rat ovary. *Histochemical Journal*, **18**, 41–4.

Gossrau, R. (1985). Cytochemistry of membrane proteases. *Histochemical Journal*, **17**, 737–71.

Gossrau, R. and Lojda, Z. (1980). Study on dipeptidylpeptidase II (DPP II). *Histochemistry*, **70**, 53–76.

Gossrau, R., Van Noorden, C. J. F., and Frederiks, W. M. (1989). Enhanced light microscopic visualization of oxidase activity with the cerium capture method. *Histochemistry*, **92**, 349–53.

Gossrau, R., Van Noorden, C. J. F., and Frederiks, W. M. (1990). Pitfalls in the light microscopical detection of NADH oxidase. *Histochemical Journal*, **22**, 155–61.

Gutschmidt, S., Kaul, W., and Riecken, E. O. (1979). A quantitative histochemical technique for the characterisation of α-glucosidases in the brush-border membrane of rat jejenum. *Histochemistry*, **63**, 81–101.

Hayhoe, F. G. J. and Quaglino, D. (1988). *Haematological Cytochemistry*, 2nd edn. Churchill Livingstone, Edinburgh.

Hiraoka, T., Hirai, K. I., and Ueda, T. (1986). Cobalt acetylacetonate-diaminobenzidine reaction for microspectrophotometry of cytochrome oxidase. *Acta Histochemica et Cytochemica*, **19**, 429–36.

Holzgreve, W., Fish, A. S., Goldsmith, P. C., and Golbus, M. S. (1985). Histochemical demonstration of ornithine transcarbamylase activity in fetal liver. *Biology of the Neonate*, **47**, 339–42.

Hopsu-Havu, V. K., Arstila, A. K., Helminen, H. J., and Kalimo, H. O. (1967). Improvements in the method for the electron microscopic localization of aryl sulphatase activity. *Histochemie*, **8**, 54–64.

Ito, S., Syundo, J., Tsuji, Y., Shimizu, I., Kishi, S., Tamura, Y., Ii, K., and Xu, Y. (1988). Cytochemical demonstration of guanase in human liver using yellow tetrazolium. *Acta Histochemica*, **83**, 99–105.

Johnson, M. A. (1991). Applications of histochemistry in muscle pathology. In *Histochemistry. Theoretical and applied*, Vol. 3 (4th edn) (ed. P. J. Stoward and A. G. E Pearse), pp. 489–514. Churchill Livingstone, Edinburgh.

Jonges, G. N. and Van Noorden, C. J. F. (1989). In situ kinetic parameters of glucose-6-phosphate dehydrogenase and phosphogluconate dehydrogenase in different areas of the rat liver acinus. *Histochemical Journal*, **21**, 585–94.

Jonges, G. N., Van Noorden, C. J. F., and Gossrau, R. (1990). Quantitative histochemical analysis of glucose-6-phosphatase activity in rat liver using an optimized cerium-diaminobenzidine method. *Journal of Histochemistry and Cytochemistry*, **38**, 1413–19.

Kaplow, L. S., Dauber, H., and Lerner, E. (1976). Assessment of monocyte esterase activity by flow cytophotometry. *Journal of Histochemistry and Cytochemistry*, **24**, 363–72.

Karnovsky, M. J. and Roots, L. (1964). A direct coloring thiocholine method for cholinesterases. *Journal of Histochemistry and Cytochemistry*, **12**, 219–21.

Klimek, F., Moore, M. A., Schneider, E., and Bannasch, P. (1988). Histochemical and microbiochemical demonstration of reduced pyruvate kinase activity in thioacetamide-induced neoplastic nodules of rat liver. *Histochemistry*, **90**, 37–42.

Kooij, A., Frederiks, W. M., Gossrau, R., and Van Noorden, C. J. F. (1991). Localization of xanthine oxidoreductase activity using the tissue protectant polyvinyl alcohol and final electron acceptor tetranitro BT. *Journal of Histochemistry and Cytochemistry*, **39**, 87–93.

Kugler, P. (1982*a*). Quantitative histochemistry of the lysosomal dipeptidyl aminopeptidase II in the proximal tubules of the rat kidney. *Histochemistry*, **76**, 557–66.

Kugler, P. (1982*b*). Quantitative dehydrogenase histochemistry with exogenous electron carriers (PMS, MPMS, MB). *Histochemistry*, **75**, 99–112.

Kugler, P. (1987). Improvement of the method of Karnovsky and Roots for the histochemical demonstration of acetylcholinesterase. *Histochemistry*, **86**, 531–2.

Kugler, P. (1990*a*). Microphotometric determination of enzymes in brain sections. III. Glutamate dehydrogenase. *Histochemistry*, **93**, 537–40.

Kugler, P. (1990*b*). Microphotometric determination of enzymes in brain sections. I. Hexokinase. *Histochemistry*, **93**, 295–8.

Kugler, P. (1991). Microphotometric determination of enzymes in brain sections. V. Glycerophosphate dehydrogenase. *Histochemistry*, **95**, 579–83.

Kugler, P. and Baier, G. (1990). Microphotometric determination of enzymes in brain sections. II. GABA transaminase. *Histochemistry*, **93**, 501–5.

Kugler, P. and Vogel, S. (1991). Microphotometric determination of enzymes in brain sections. IV. Isocitrate dehydrogenase. *Histochemistry*, **95**, 629–33.

Kugler, P., Vogel, S., and Gehm, M. (1988*a*). Quantitative succinate dehydrogenase histochemistry in the hippocampus of aged rats. *Histochemistry*, **88**, 299–308.

Kugler, P., Vogel, S., Volk, H., and Schiebler, T. H. (1988*b*). Cytochrome oxidase histochemistry in the rat hippocampus. A quantitative methodological study. *Histochemistry*, **89**, 269–75.

Lawrence, G. M., Beesley, A. C. H., and Matthews, J. B. (1989). The use of continuous monitoring and computer-assisted image analysis for the histochemical quantification of hexokinase activity. *Histochemical Journal*, **21**, 557–64.

Lojda, Z., Gossrau, R., and Schiebler, T. H. (1979). *Enzyme Histochemistry. A Laboratory Manual*. Springer Verlag, New York.

Lund-Hansen, T., Hoyer, P. E., and Anderson, M. (1984). A quantitative cytochemical assay of β-galactosidase in single cultured human skin fibroblasts. *Histochemistry*, **81**, 321–30.

Macha, N., Bitensky, L., and Chayen, J. (1975). The effect of thyrotrophin on oxidative activity in thyroid follicle cells. *Histochemistry*, **41**, 323–34.

Martin, T. P., Vailas, A. C., Durivage, J. B., Edgerton, V. R., and Castleman, K. R. (1985). Quantitative histochemical determination of muscle enzymes: biochemical verification. *Journal of Histochemistry and Cytochemistry*, **33**, 1053–9.

Mayer, D., Ehemann, V., Hacker, H. J., Klimek, F., and Bannasch, P. (1985). Specificity of cytochemical demonstration of adenylate cyclase in liver using adenylate-$(\beta,\gamma$-methylene) diphosphate as substrate. *Histochemistry*, **82**, 135–40.

McGarry, A. and Gahan, P. B. (1985). A quantitative cytochemical study of UDP-D-glucose: NAD-oxidoreductase (E.C. 1.1.1.22) activity during stellar differentiation in *Pisum sativum* L. cv Meteor. *Histochemistry*, **83**, 551–4.

McMillan, P. J. (1967). Differential demonstration of muscle and heart type lactic dehydrogenase of rat muscle and kidney. *Journal of Histochemistry and Cytochemistry*, **15**, 21–31.

Meijer, A. E. F. H. (1972). Semipermeable membranes for improving the histochemical demonstration of enzyme activities in tissue sections. I Acid phosphatase. *Histochemie*, **30**, 31–9.

Meijer, A. E. F. H. (1980). Semi-permeable membrane techniques in quantitative enzyme histochemistry. In *Trends in Enzyme Histochemistry and Cytochemistry* (ed. D. Evered and M. O'Connor), pp. 103–20. Excerpta Medica, Amsterdam.

Mizukami, Y., Matsubara, F., Matsukawa, S. and Izumi, R. (1983). Cytochemical localization of glutaraldehyde-resistant NAD(P)H oxidase in rat hepatocytes. *Histochemistry*, **79**, 259–67.

Patel, H. R. H., Frederiks, W. M., Marx, F., Best, A. J., and Van Noorden C. J. F. (1991). A quantitative histochemical study of D-amino acid oxidase activity in rat liver in relationship with feeding conditions. *Journal of Histochemistry and Cytochemsitry*, **39**, 81–6.

Papadimitriou, J. M. and Van Duijn, P. (1970). Effects of fixation and substrate protection on the isoenzymes of aspartate aminotransferase studied in a quantitative cytochemical model system. *Journal of Cell Biology*, **47**, 71–83.

Rieder, H. (1981). NADP-dependent dehydrogenases in rat liver parenchyma. III. The description of a liponeogenic area on the basis of histochemically demonstrated enzyme activities and the neutral fat content during fasting and refeeding. *Histochemistry*, **72**, 579–615.

Rieder, H., Teutsch, H. F., and Sasse, D. (1978). NADP-dependent dehydrogenases in rat liver parenchyma. I. Methodological studies on the qualitative histochemistry of G6PDH, 6PGDH, malic enzyme and ICDH. *Histochemistry*, **56**, 283–98.

Ritter, J. (1973). Quantitative Untersuchungen zum histochemischen Nachweis von GABA-Transaminase-SSA-Dehydrogenase. *Acta Histochemica*, **47**, 153–75.

Robertson, W. R. (1979). A quantitative cytochemical method for the demonstration of Δ^5 - 3β-hydroxysteroid dehydrogenase activity in unfixed tissue sections of rat ovary. *Histochemistry*, **59**, 271–85.

Robertson, W. R. (1980). A quantitative study of *N*-acetyl-β-glucosaminidase activity in unfixed tissue sections of the guinea pig thyroid gland. *Histochemical Journal*, **12**, 87–96.

Robertson, W. R., Frost, J., Hoyer, P. E., and Weinkove, C. (1982). 20α-Hydroxysteroid dehydrogenase activity in the rat corpus luteum; a quantitative cytochemical study. *Journal of Steroid Biochemistry*, **17**, 237–43.

Roels, F., Wisse, E., DePrest, B., and Van der Meulen, J. (1975). Cytochemical discrimination between catalases and peroxidases using diaminobenzidine. *Histochemistry*, **41**, 281–312.

Rudolphus, A., Van Twisk, C., Van Noorden, C. J. F., Dijkman, J. H., and Kramps, J. A. (1992). Detection of extracellular neutrophil elastase in hamster lungs after intratracheal instillation of *E. coli* lipopolysaccharide using a fluorogenic, elastase specific, synthetic substrate. *American Journal of Pathology*, in press.

Schofield, K. P., Stone, P. C. W., Beddall, A. C., and Stuart, J. (1983). Quantitative cytochemistry of the toxic granulation blood neutrophil. *British Journal of Haematology*, **53**, 15–22.

Stoward, P. J. and Van Noorden, C. J. F. (1991). Histochemical methods for dehydrogenases. In *Histochemistry. Theoretical and applied*, Vol. 3 (4th edn) (ed. P. J. Stoward and A. G. E. Pearse), pp. 537–57. Churchill Livingstone, Edinburgh.

Straatsburg, I. H., De Graaf, F., Van Noorden C. J. F., and Van Raamsdonk, W. (1989). Enzyme reaction rate studies in electromotor neurons of the weakly electric fish *Apteronotus leptorhynchus*. *Histochemical Journal*, **21**, 609–17.

Stuart, J. and Simpson, J. S. (1970) Dehydrogenase enzyme cytochemistry of unfixed leukocytes. *Journal of Clinical Pathology*, **23**, 517–21.

Van Goor, H. and Hardonk, M. J. (1990). Cerium techniques in light microscopy: a short review. *Transactions of the Royal Microscopical Society*, **1**, 565–8.

Van Goor, H., Poelstra, K., and Hardonk, M. J. (1989). Cerium-based demonstration of phosphatase activity in plastic-embedded sections: A comparison with conventional methods. *Stain Technology*, **64**, 289–96.

Van Noorden, C. J. F. (1984). Histochemistry and cytochemistry of glucose-6-phosphate dehydrogenase. *Progress in Histochemistry and Cytochemistry*, **15**(4), 1–85.

Van Noorden, C. J. F. and Butcher, R. G. (1984). Histochemical localization of NADP-dependent dehydrogenase activity with four different tetrazolium salts. *Journal of Histochemistry and Cytochemistry*, **32**, 998–1004.

Van Noorden, C. J. F. and Butcher, R. G. (1986). A quantitative histochemical study of NADPH-ferrihemoprotein reductase activity. *Histochemical Journal*, **18**, 364–70.

Van Noorden, C. J. F. and Butcher, R. G. (1991). Quantitative enzyme histochemistry. In *Histochemistry. Theoretical and Applied*, Vol. 3, 4th edn. (ed. P. J. Stoward and A. G. E. Pearse), pp. 355–432. Churchill Livingstone, Edinburgh.

Van Noorden, C. J. F. and Jonges, G. N. (1987). Quantification of the histochemical reaction for alkaline phosphatase activity using the indoxyl-tetranitro BT method. *Histochemical Journal*, **19**, 94–102.

Van Noorden, C. J. F. and Tas, J. (1982). The role of exogenous electron carriers in

NAD(P)-dependent dehydrogenase cytochemistry studied *in vitro* and with a model system of polyacrylamide films. *Journal of Histochemistry and Cytochemistry*, **30**, 12–20.

Van Noorden, C. J. F. and Vogels, I. M. C. (1989*a*). Polyvinyl alcohol and other tissue protectants in enzyme histochemistry: a consumer's guide. *Histochemical Journal*, **21**, 373–9.

Van Noorden, C. J. F. and Vogels, I. M. C. (1989*b*). Cytophotometric analysis of reaction rates of succinate and lactate dehydrogenase activity in rat liver, heart muscle and tracheal epithelium. *Histochemical Journal*, **21**, 575–83.

Van Noorden, C. J. F., Vogels, I. M. C., Everts, V., and Beertsen, W. (1987). Localization of cathepsin B activity in fibroblasts and chondrocytes by continuous monitoring of the formation of a final fluorescent reaction product using 5'-nitrosalicylaldehyde. *Histochemical Journal*, **19**, 483–7.

Van Noorden, C. J. F., Vogels, I. M. C., and Van Wering, E. R. (1989*a*). Enzyme cytochemistry of unfixed leukocytes and bone marrow cells using polyvinyl alcohol for the diagnosis of leukemia. *Histochemistry*, **92**, 313–8.

Van Noorden, C. J. F., Vogels, I. M. C., and Smith, R. E. (1989*b*). Localization and cytophotometric analysis of cathepsin B activity in unfixed and undecalcified cryostat sections of whole rat knee joints. *Journal of Histochemistry and Cytochemistry*, **37**, 617–24.

Van Reempts, J., Van Deuren, B., Haseldonkx, M., Van de Ven, M., Thoné, F., and Borgers, M. (1988). Purine nucleoside phosphorylase: a histochemical marker for glial cells. *Brain Research*, **462**, 142–7.

Wachsmuth, E. D. and Donner, P. (1976). Conclusions about aminopeptidase in tissue sections from studies of amino acid naphthylamide hydrolysis. *Histochemistry*, **47**, 271–83.

Watanabe, J., Kanai, K., and Kanamura, S. (1989). A new microphotometric method for measurement of cytochrome P-450 in sections of liver. *Journal of Histochemistry and Cytochemistry*, **37**, 1257–63.

Whittaker, J. R. (1981). Quantitative measurement by microdensitometry of tyrosinase (DOPA oxidase) development in whole small ascidian embryos. *Histochemistry*, **71**, 349–59.

Wimmer, M. and Pette, D. (1979). Microphotometric studies on intraacinar enzyme distribution in rat liver. *Histochemistry*, **64**, 23–33.

5 Methods developed for clinical and experimental pathology

Enzyme histochemical methods are used routinely, particularly for the diagnosis of leukaemia (for a review, see Hayhoe and Quaglino 1988) and of muscle diseases (for a review, see Johnson 1991). These methods are not described any further here, because they are similar to those in Chapter 4.

A number of enzyme histochemical methods have, however, been designed to answer specific questions in clinical and experimental pathology, such as the oxygen sensitivity test to detect malignant and potentially malignant tumour cells, the 'nothing' dehydrogenase reaction for the detection of myocardial infarction, the lysosomal membrane fragility test, the use of unfixed and undecalcified bone sections, and the assay for the detection of heterozygous glucose-6-phosphate dehydrogenase deficiency in erythrocytes.

5.1 Oxygen sensitivity test for the diagnosis of malignancy

The metabolic patterns of cancer cells often differ from those of normal cells long before morphological changes are manifest. In principle, the altered metabolism enables early diagnosis of malignancy. For example, cancer cells are glycolytic and show high glucose-6-phosphate dehydrogenase activity. The latter enzyme has been used in the past as a discriminator between normal and cancer cells. However, because high activity may be present in any proliferating cell, malignant, benign or normal, it is not sufficient simply to demonstrate glucose-6-phosphate dehydrogenase activity in order to diagnose malignancy (Van Noorden 1984). By using an oxygen-sensitive tetrazolium salt (neotetrazolium chloride) for the histochemical demonstration of glucose-6-phosphate dehydrogenase activity, it is possible to distinguish malignant cells from normal cells. Normal cells do not contain any final reaction product, formazan, after being incubated for 5–10 min at 37 °C in the presence of neotetrazolium chloride and oxygen for the demonstration of glucose-6-phosphate dehydrogenase activity, whereas malignant cells do. When oxygen is replaced by nitrogen to create anaerobic conditions, normal and malignant cells show their activity. The principle of this discriminating capacity of the neotetrazolium test reaction in oxygen is probably based on decreased superoxide dismutase activity in malignant cells. This enzyme interferes with the reduction of neotetrazolium which takes place via the formation of oxygen radicals when the reaction is performed under aerobic conditions (Best et al. 1990). The method is described in Method 48.

Method 48. *Oxygen sensitive tetrazolium salt method for the detection of glucose-6-phosphate dehydrogenase activity in normal and malignant cells (Best et al. 1990)*

1. Prepare incubation medium for the histochemical demonstration of glucose-6-phosphate dehydrogenase activity according to Method 33, but replace 5 mM tetranitro BT by 5 mM neotetrazolium chloride (Polysciences, Northampton, UK). When neotetrazolium is used from other suppliers, it may be contaminated with impurities and needs purification (Altman 1976).
2. Divide medium into two equal parts and saturate one part with 100 per cent oxygen and the other with 100 per cent nitrogen. Because of the viscosity of polyvinyl alcohol-containing media, use a tonometer for gassing the media (Butcher 1978). When a tonometer is not available, the media have to stand in air-tight vials for 15–30 min after the gas has been bubbled through, in order to eliminate bubbles from the medium.
3. Use unfixed cryostat sections of the tissue to be diagnosed (8–10 μm thick).
4. Pour the media on to the sections and incubate for 10 min at 37 °C.
5. Rinse off media with 100 mM phosphate buffer (pH 5.3) at 60 °C.
6. Mount sections in glycerine–gelatin.
7. Analyse the sections: areas with malignant cells contain substantial amounts of formazan after incubation in oxygen whereas normal cells do not. Use the sections incubated in nitrogen as a reference.

5.2 'Nothing' dehydrogenase reaction for the detection of necrosis by ischaemia

The size of infarcted areas in heart can be determined by incubating thick slices in a tetrazolium salt solution. Formazan is generated in normal areas whereas necrotic areas remain unstained (Sandritter and Jestadt 1958). Identification of necrosis induced by ischaemia has also been performed in a similar way in skeletal muscle, brain, and liver tissue (for an overview, see Frederiks et al. 1989). The formazan production depends on the presence of dehydrogenases (mainly lactate dehydrogenase), dehydrogenase substrates, and/or thiol-containing proteins (Van Noorden et al. 1985). This reaction is called the 'nothing' dehydrogenase reaction because the incubation medium lacks a specific substrate. Ischaemia without reperfusion reduces the 'nothing' dehydrogenase reaction mainly because of degradation of substrates and thiol groups whereas ischaemia followed by reperfusion leads to loss of lactate dehydrogenase as well (see Method 49).

Method 49. *'Nothing' dehydrogenase reaction for detection of necrosis (Frederiks et al. 1989)*

1. Prepare fresh tissue slices (2–10 mm thick) or cryostat sections (8–10 µm thick).
2. Incubate serial slices or sections for 5 min at 37 °C in medium containing 100 mM Tris–maleate buffer (pH 7.45), 18 per cent polyvinyl alcohol (see Method 14), 5 mM tetranitro BT, 5 mM sodium azide, 0.2 mM phenazine methosulfate, with or without 3 mM NAD^+.
3. Rinse slices or sections with 100 mM phosphate buffer (pH 5.3) at 60 °C.
4. Mount sections in glycerine–gelatin.
5. Determine areas that are unstained after being incubated in the presence and absence of NAD^+. These areas have lost lactate dehydrogenase activity, indicating irreversible damage due to necrosis. Areas that stain in the presence of NAD^+ but not in the absence of NAD^+ are damaged reversibly by loss of endogenous substrates and thiol groups. The areas stained in the absence of NAD^+ may be considered as unaffected.

5.3 Lysosomal membrane fragility test

The lysosomal membrane fragility test is a quantitative cytochemical approach for detecting cellular damage. Toxic compounds or pathophysiological conditions often have an effect on lysosomes by changing either the content of hydrolytic enzymes or the state of the lysosomal membrane. The latter aspect can be tested by the membrane fragility test. The principle of the test is simple: an unfixed cryostat section or cell preparation is incubated for the demonstration of the activity of a lysosomal enzyme such as acid phosphatase or β-glucuronidase. In the first phase of the incubation, lysosomal membranes are impermeable to compounds in the reaction medium and no final reaction product is formed. During incubation at low pH, the membranes are modified and become permeable and so reaction product is generated. The initial period of latency (or lag phase) is an indicator of the stability of the membranes. The longer the lag phase, the better the condition of the cell or tissue (Bitensky *et al.* 1973; Moore 1976). The duration of the lag phase can be detected semi-quantitatively by monitoring the period of time necessary before final reaction product starts to occur in a simultaneous diazonium azo-coupling method (see Methods 28 and 29). However, continuous monitoring of the formation of final reaction product of the simultaneous azo-coupling method using a cytophotometer or an image analysing system is a more accurate and reliable approach (see Sections 2.4 and 2.5). The method has been used successfully for the analysis of cellular damage due to inflammatory or arthritic processes and in the study of effects of xenobiotics on cells due to e.g. marine pollution (for a brief review, see Van Noorden 1991).

When tissue protectants such as polyvinyl alcohol (see Method 14) are used in the incubation medium in order to improve morphology of the tissue, the lag phase is lengthened due to the stabilizing effects of tissue protectants (Bitensky et al. 1973). A prolonged lag phase enables a better discrimination of control and affected cells and therefore, the use of polyvinyl alcohol is also recommended in the fragility test. One should also be aware of the fact that the lag phase can be shortened or circumvented experimentally either by preincubation of sections at an acid pH (e.g. in acetate buffer) or by a pretreatment such as freeze–thawing. These pretreatments can be useful when the total activity of an acid hydrolase has to be demonstrated but should be avoided in the fragility test (Chayen et al. 1973). Furthermore, not all substrates for acid hydrolases are suitable for the fragility test, e.g. 4-methoxy-2-naphthylamide derivatives penetrate lysosomal membranes without any difficulty. Hydrolase reactions in which these substrates are used do not show a distinct lag phase (Chayen et al. 1973). Diazonium salts can shorten the period of the lag phase and/or inhibit enzyme activity. Therefore, Moore (1976) prefers post-coupling methods in the fragility test. The general procedure for the lysosomal membrane fragility test is described in Method 50.

Method 50. *Lysosomal membrane fragility test (Bitensky et al. 1973; Moore 1976)*

1. Prepare unfixed cryostat sections (8–10 μm thick) as described in Method 10. Store sections in cryostat cabinet until used (30–120 min).
2. Prepare incubation media for simultaneous coupling methods to demonstrate acid phosphatase and/or β-glucuronidase activity as described in Methods 28 and 29.
3. Pour incubation media at 37 °C on to sections and monitor formation of final reaction product as described in Method 2. Avoid drying of sections prior to incubation because this step increases permeability of lysosomal membranes.
4. Establish the period of time of incubation necessary before final reaction product starts to be formed.

5.4 Unfixed and undecalcified bone sections

Ullberg (1954) developed a method for cutting whole body cryostat sections of laboratory animals without previous fixation or demineralization for autoradiography in pharmacological studies. The integrity of the large sections is preserved by applying adhesive tape to the cutting surface (see

Method 51). The method has also been used for enzyme histochemical reactions in whole body sections of mice (Egashira and Waddell 1984), oral cavity (Sjögren 1984), and knee joints (Van Noorden and Vogels 1986). All the enzymes listed in Table 2.1 can be demonstrated in these sections because the detrimental effects of fixation and decalcification are avoided.

Method 51. *Preparation of unfixed and undecalcified bone sections (Van Noorden and Vogels 1986)*

1. Embed tissue containing bone in an aqueous solution of 8 per cent (w/v) gelatin that has previously been poured into a mould.
2. Freeze the mould containing gelatin and tissue very slowly in liquid nitrogen. The gelatin and tissue crack easily if frozen too rapidly.
3. Store material as described in Section 3.4.1.
4. Before sections are cut, adjust the temperature of the block of tissue and gelatin to the temperature of the cryostat (see Section 3.2.1.1).
5. Use a tungsten carbide tipped knife (Spikker, Zevenaar, The Netherlands) in the cryostat. This knife is specially developed for cutting hard material such as bone. The angle between knife and face of the tissue block has to be 10°.
6. Trim the tissue block to the desired level.
7. Firmly fasten transparent Scotch tape 800 (3M, St Paul, MN, USA) onto the surface of the block using a brush.
8. Cut a section underneath the tape at an extremely low but constant speed.
9. Fix the piece of tape with its adhering section to a glass slide using ordinary tape.
10. (Optional) Check morphological aspect of section by staining briefly using a Methylene Blue or Giemsa solution as described in Section 3.2.1.1.
11. Perform enzyme incubations as described in the methods of Chapter 4 and rinse section thoroughly.
12. Cut out the section with a pair of scissors and mount this section with underlying tape in a double layer of glycerol jelly.

5.5 Detection of glucose-6-phosphate dehydrogenase deficiency in erythrocytes

The most common inherited enzyme deficiency in man is the X-linked deficiency in glucose-6-phosphate dehydrogenase. The main clinical manifestation of the disease is haemolysis as erythrocytes are the only cells seriously affected. At least 40 per cent of heterozygote female patients cannot be diagnosed on the basis of spectrophotometric analysis of haemolysates and a reliable cytochemical method is thus an important diagnostic tool by demonstrating the intercellular distribution of the enzyme activity in erythrocytes. The relatively low glucose-6-phosphate dehydrogenase activity in each erythrocyte requires a highly sensitive cytochemical method for reliable diagnosis. This has been achieved by combining very mild fixation of the cells with concomitant protection of the enzymic active site by NADP$^+$ (see Section 2.6.12) to permeabilize the erythrocyte membrane, using polyvinyl alcohol to reduce diffusion of intermediate reaction products, a high incubation temperature to speed up the enzyme reaction, and prolonged incubation periods (Van Noorden and Vogels 1985). This technique, as presented in Method 52, enables the detection of deficient cells in a mixed population from heterozygously deficient patients by either microscopic analysis (Vogels *et al.* 1986) or flow cytofluorometric analysis (Van Noorden *et al.* 1989).

Method 52. *Localization of glucose-6-phosphate dehydrogenase activity in erythrocytes (Van Noorden and Vogels 1985)*

1. Incubate 1 ml of whole blood in 9 ml of a freshly prepared solution of sodium nitrite (180 mM) in 0.9 per cent sodium chloride for 8 min at room temperature for the conversion of all oxyhaemoglobin into methaemoglobin to prevent aspecific formazan production.
2. Centrifuge cell suspension for 15 min at 1000g at room temperature.
3. Add 75 µl of the packed cells to 125 µl of a 20 mM solution of NADP$^+$ in 100 mM phosphate buffer (pH 7.0) and incubate for 10 min at room temperature.
4. Centrifuge cells for 15 min at room temperature at 1000g.
5. Pour off supernatant and add 2 ml of a solution containing 0.025 per cent (v/v) depolymerized glutaraldehyde in 100 mM phosphate buffer (pH 7.0) to the packed cells. The relative amounts of cells and glutaraldehyde in the mixture is very critical. Too many cells results in improper permeabilization of the cells, whereas too few cells results in inactivation of glucose-6-phosphate dehydrogenase.
6. Incubate the mixture for exactly 30 min at room temperature with continuous rotation of the tubes to maximize interaction between cells and fixative.
7. Wash cells three times with 100 mM phosphate buffer (pH 7.0) at 4 °C, for 1 min each time followed by rapid centrifugation at 2000g.
8. Resuspend packed cells in 0.5 ml of the same buffer and add 100 µl of this cell suspension to 1 ml of the incubation medium used to demonstrate glucose-6-phosphate dehydrogenase activity (prepared according to Method 33). The medium consists of 100 mM phosphate buffer (pH 7.0), 20 per cent polyvinyl alcohol (see Method 14), 10 mM glucose-6-phosphate, 0.8 mM NADP$^+$, 0.45 mM 1-methoxyphenazine methosulfate, 5 mM sodium azide, and 1 mM tetranitro BT.
9. Incubate for 90 min at 46 °C in the dark, with constant rotation.
10. Stop reaction by adding ice-cold phosphate buffer (pH 7.0).
11. Wash cells three times with the same buffer at 4 °C, for 1 min each time followed by rapid centrifugation at 2000g.
12. Resuspend cells in 1 ml buffer and place 3 µl of the suspension on to a glass slide.
13. Dry cells in air for 30 min at 37 °C and mount cells in water-free immersion glycerol for microscopical inspection (Vogels *et al.* 1986) or analyse cell

suspension quantitatively by flow cytofluorometry (Van Noorden et al. 1989).
14. For microscopy, analyse 1000 cells to determine whether or not they contain formazan. A person should be considered as normal when < 20 per cent negative cells are found and as heterozygously deficient when between 20 and 80 per cent of cells are negative. When > 80 per cent negative cells are found, consider males as hemizygously deficient and females as homozygously deficient.

References

Altman, F. P. (1976). Tetrazolium salts and formazans. *Progress in Histochemistry and Cytochemistry*, **9**(3), 1–56.
Best, A. J., Das, P. K., Patel, H. R. H., and Van Noorden, C. J. F. (1990). Quantitative cytochemical detection of malignant and potentially malignant cells in the colon. *Cancer Research*, **50**, 5112–8.
Bitensky, L., Butcher, R. G., and Chayen, J. (1973). Quantitative cytochemistry in the study of lysosomal function. In *Lysosomes in Biology and Pathology* (ed. J. T. Dingle), pp. 465–510. North-Holland, Amsterdam.
Butcher, R. G. (1978). Oxygen and the production of formazan from neotetrazolium chloride. *Histochemistry*, **56**, 329–40.
Chayen, J., Bitensky, L., and Butcher, R. G. (1973). *Practical Histochemistry*. Wiley, London.
Egashira, T. and Waddell, W. J. (1984). Histochemical localization of primary and secondary alcohol dehydrogenases in whole-body, freeze-dried sections of mice. *Histochemical Journal*, **16**, 931–40.
Frederiks, W. M., Marx, F., and Myagkaya, G. L. (1989). The 'nothing dehydrogenase' reaction and the detection of ischaemic damage. *Histochemical Journal*, **21**, 565–73.
Hayhoe, F. G. J. and Quaglino, D. (1988). *Haematological Cytochemistry*, 2nd edn. Churchill Livingstone, Edinburgh.
Johnson, M. A. (1991). Applications of enzyme histochemistry in muscle pathology. In *Histochemistry. Theoretical and Applied*, Vol. 3 (4th edn) (ed. P. J. Stoward and A. G. E. Pearse), pp. 489–514. Churchill Livingstone, Edinburgh.
Moore, M. N. (1976). Cytochemical demonstration of latency of lysosomal hydrolases in digestive cells of the common mussel, *Mytilus edulis*, and changes induced by thermal stress. *Cell and Tissue Research*, **175**, 279–87.
Sandritter, H. W. and Jestadt, R. (1958). Triphenyltetrazolium Chlorid (TTC) als Reduktionsindikator zur makroskopischen Diagnose des frischen Herzinfarktes. *Verhandlungen der Deutschen Gesellshaft für Pathologie*, **41**, 165–70.
Sjögren, S. (1984). Lactate dehydrogenase in developing rat oral epithelium. *Journal of Histochemistry and Cytochemistry*, **32**, 1–6.
Ullberg, S. (1954). Studies on the distribution and fate of S^{35} labeled benzyl-penicillin in the body. *Acta Radiologica, Supplement* **118**, 1–135.
Van Noorden, C. J. F. (1984). Histochemistry and cytochemistry of glucose-6-

phosphate dehydrogenase. *Progress in Histochemistry and Cytochemistry*, **15**(4), 1–84.

Van Noorden, C. J. F. (1991). Assessment of lysosomal function by quantitative histochemical and cytochemical methods. *Histochemical Journal*, **23**, 429–35.

Van Noorden, C. J. F. and Vogels, I. M. C. (1985). A sensitive cytochemical staining method for glucose-6-phosphate dehydrogenase activity in individual erythrocytes. II. Further improvements of the staining procedure and some observations with glucose-6-phosphate dehydrogenase deficiency. *British Journal of Haematology*, **60**, 57–63.

Van Noorden, C. J. F. and Vogels I. M. C. (1986). Enzyme histochemical reactions in unfixed and undecalcified cryostat sections of mouse knee joints with special reference to arthritic lesions. *Histochemistry*, **86**, 127–33.

Van Noorden, C. J. F., Kooij, A., Vogels, I. M. C., and Frederiks, W. M. (1985). On the nature of the 'nothing dehydrogenase' reaction. *Histochemical Journal*, **17**, 1111–8.

Van Noorden, C. J. F., Dolbeare, F., and Aten, J. (1989). Flow cytofluorometric analysis of enzyme reactions based on quenching of fluorescence by the final reaction product: detection of glucose-6-phosphate dehydrogenase deficiency in human erythrocytes. *Journal of Histochemistry and Cytochemistry*, **37**, 1313–18.

Vogels, I. M. C., Van Noorden, C. J. F., Wolf, B. H. M., Saelman, D. E. M., Tromp, A., Schutgens R. B. H., and Weening R. S. (1986). Cytochemical determination of heterozygous glucose-6-phosphate dehydrogenase deficiency in erythrocytes. *British Journal of Haematology*, **63**, 402–5.

Index

acetyl esterases 67, 68
  method 65, 68
  see also non-specific esterases
acetylcholinesterases 24, 67
  enzyme classification 5
  method 92
  properties 91, 92
acid α-glucosidase 65
  enzyme classification 5
  method 65
  properties 65–6
acid phosphatase
  enzyme classification 5
  method 64–5
  properties 64
  see also lysosomal membrane fragility test
adenylate cyclase
  enzyme classification 5
  method 52, 55
  properties 55
affinity histochemistry 1
aldehyde dehydrogenase
  enzyme classification 4
  method 72–3, 77
  properties 77
aldolase, see fructose–biphosphate aldolase
alkaline phosphatase 17, 19, 23, 60
  enzyme classification 5
  method 87–8
  properties 87
γ-aminobutyric acid transaminase
  enzyme classification 5
  method 80, 81
  properties 81
aminopeptidase A
  enzyme classification 5
  method 68, 69
  properties 69
aminopeptidase M
  enzyme classification 5
  method 68, 69
  properties 69
arthritis, see cellular damage
aryl esterases 67
  method 65, 68
  properties 68
  see also non-specific esterases
arylsulfatase 52
  enzyme classification 5
  method 59
  properties 59
aspartate aminotransferase 52
  enzyme classification 5
  method 58–9
  properties 58
ATPases 52, 53, 60
  enzyme classification 5
  Ca^{2+}-myosin ATPase properties 54
  method 55–6
  Na^+,K^+-ATPase properties 54
  method 55–6
auxiliary enzyme(s) 22, 29, 72, 79, 80
  see also multi-step reaction
azo-dye 20, 21

benzaldehyde dehydrogenase
  enzyme classification 4
  method 72–3, 78
  properties 78
bioluminescence 29

Ca^{2+}-myosin ATPase, see ATPases
carboxyl esterases
  method 65, 68
  properties 68
  see also non-specific esterases
catalase 23
  enzyme classification 4
  method 88–9
  properties 88
cathepsin B 21
  enzyme classification 5
  fluorescence method 91
  method 70–1
  properties 70, 90
cell preparations 16, 36–7, 43–6, 48, 50, 104
  fixed
    method 44
  snap freezing 36–7
  unfixed 8, 25, 36, 43, 44
    method 44
cellular damage 104–5
  see also lysosomal membrane fragility test
cerium–diaminobenzidine–hydrogen peroxide procedure 60
cerium–diaminobenzidine–cobalt–hydrogen peroxide procedure 62, 63
cerium salt capture method 20, 59–60, 61

Index

cholinesterases 19, 20, 21, 24, 52, 67
coated glass slides 42
  method 42
control reaction 2, 7, 8, 10, 13, 14, 15, 16, 18
  see also quantitative histochemistry
creatine kinase 22, 80
  enzyme classification 5
  method 80, 82, 83
  properties 82–3
cryostat section 51
  fixed 39, 42, 52
    material 54
  freeze-dried 27
  undecalcified bone 102, 105, 106
    method 106
  unfixed 8, 11, 16, 25, 36, 40, 42, 46, 48, 49, 50, 51, 52, 104
    preparation 36, 40
    ultrastructure 26
  whole body 106
cutting sections 39–42
  cryostat 17, 27, 40
  fixed cryostat sections 42
  glycol methacrylate sections 42
  unfixed cryostat sections 40
cysteine proteinase 21, 68
  see also cathepsin B
cytochrome c oxidase 88
  enzyme classification 4
  method 90
  properties 89
cytochrome P450 25
  method 95
  properties 94
cytophotometry 1, 8, 11, 14, 15, 18, 27, 28, 29
  arbitrary machine units 29
  cytophotometer 16, 17, 18, 27, 28, 104
  distributional errors 15
  mean integrated absorbance 12, 17, 18, 28, 29
cytospin preparations, see cell preparations

D-amino acid oxidase 51
  enzyme classification 4
  method 62
  properties 61
D,T-diaphorase, see NAD(P)H dehydrogenase
dehydrogenases 17, 21, 22, 72, 79, 103
diagnosis of leukaemia 67, 102
diaminobenzidine methods 23, 88, 89, 90
diazonium salt methods 17, 20–1, 64–71, 104–5
  post-coupling 17, 21, 68, 70, 105
    method 64–5
  simultaneous coupling 17, 21, 64, 65, 67, 104
    method 68–9
diffusion, see precision

dipeptidyl peptidase II
  enzyme classification 5
  method 68–9, 70
  properties 70
dipeptidyl peptidase IV 70
  enzyme classification 5
  method 68–9, 70
  properties 70
disaccharidases 65
  see also glycosidases
distributional errors 15
  see also cytophotometry
DOPA-oxidase, see monophenol monoxygenase

elastase 90
  enzyme classification 5
  method 90, 91
  properties 90
electron microscopical enzyme histochemistry
  cerium salt capture methods 20, 62
  diaminobenzidine methods 23
  metal salt capture methods 19–20
  unfixed cryostat sections 25–6
embedding media, see glycol methacrylate
end-point measurements 16, 17, 19
  see also quantitative enzyme histochemistry
endopeptidases 68
enzyme classification 3–6
erythrocytes 106, 107
esterases 20, 23, 65, 67–8
  method 65
  see also 'non-specific' esterases
exopeptidases 68

fixation (chemical) 10, 11, 25, 26, 48, 88, 106, 107
  fixation and embedding 38–9
  fixation and snap freezing 37
  immersion fixation of organs 38
  perfusion fixation of organs 38
  perfusion fixation of whole animals 37
flow cyto(fluoro)metry, see quantitative histochemistry
fluorescence method 23–4, 90–1
formazans 22, 23
  isobestic wavelengths 27, 28
  molecular extinction coefficient 27
freeze-substitution 36, 39
fructose–biphosphate aldolase 22
  enzyme classification 5
  method 80, 83
  properties 83

α-galactosidase 66
  enzyme classification 5

method 65, 66
properties 66
ß-galactosidase
  enzyme classification 5
  method 72
  properties 71
glucose-6-phosphatase
  enzyme classification 5
  kinetic parameters 7–9, 30
  metabolic flux rate 30
  method 60
  properties 60
  units of enzyme activity 29
glucose-6-phosphate dehydrogenase
  cytophotometric analysis 28
  deficiency in erythrocytes 106–7
    method 107
  enzyme classification 4
  localization in liver 8
  method 72–3, 76
  oxygen sensitivity test 102–3
    method 103
  properties 76
glucose-6-phosphate isomerase 83
  enzyme classification 6
  method 80, 83–4
  properties 83–4
ß-glucuronidase
  enzyme classification 5
  method 65, 67
  properties 67
  see also lysosomal membrane fragility test
glutamate dehydrogenase
  enzyme classification 4
  immunohistochemistry 32
  in situ hybridization 32
  localization in blood cells 45
  localization in liver 32
  method 72–3, 79
  properties 79
γ-glutamyltransferase
  enzyme classification 5
  method 68–9
  properties 69
glyceraldehyde-3-phosphate dehydrogenase
  enzyme classification 4
  method 72–3, 78
  properties 78
glycerol-3-phosphate dehydrogenase
  enzyme classification 4
  method 72–3, 77
  properties 77
glycerol-3-phosphate dehydrogenase (NAD$^+$-dependent)
  enzyme classification 3
  method 72–3
  properties 73
glycogen 24
  see also glycogen phosphorylase

glycogen phosphorylase 24, 50
  enzyme classification 5
  method 93
  properties 92–3
glycol methacrylate 36, 38, 39, 42, 46, 48
glycosidases 20, 21, 23, 64, 65
glycosyltransferases 24
guanine deaminase 83
  enzyme classification 5
  method 80, 83
  properties 83

Hatchett brown 24
hexokinase 22
  enzyme classification 5
  kinetic parameters 14
  method 80, 81
  properties 81
hydrolases 5, 50
α-hydroxy acid oxidase 61
  enzyme classification 4
  method 62–3
  properties 62
20α-hydroxysteroid dehydrogenase
  enzyme classification 4
  method 72–3, 76
  properties 76
3ß-hydroxy-Δ^5-steroid dehydrogenase
  enzyme classification 4
  method 72–3, 77
  properties 77
3-hydroxyacyl CoA dehydrogenase
  enzyme classification 4
  method 72–3, 77
  properties 74
3-hydroxybutyrate dehydrogenase
  enzyme classification 4
  method 72–3, 74
  properties 74

image analysis, see quantitative histochemistry
immunohistochemistry 1, 23, 30–2
in situ hybridization 1, 30–2
incubation conditions
  aqueous media 48, 50, 52
  polyvinyl alcohol 11, 14, 25, 43, 44, 48–50, 52, 67, 105, 107
    method 49
  semipermeable membrane technique 11, 25, 26, 48, 50–2, 61, 92, 93
    method 51–2
indigogenic methods 21, 71–2
indoxyl-tetrazolium salt method 21, 23, 60, 87–8
inflammation, see cellular damage
inhibitors 8, 12, 13, 15, 16, 18, 22, 31, 67, 84, 88

inhibitors (*cont.*)
  allosteric 9
  competitive 9, 22
  inhibitor constant 9
  non-competitive 9
  specific 52
  see also kinetic parameters
inter-assay variation 13
  see also quantitative histochemistry
intermediate reaction products 49
intra-assay variation 13
  see also quantitative histochemistry
ischaemia, *see* necrosis
isobestic wavelength 27–8
isocitrate dehydrogenase (NAD$^+$-dependent)
  enzyme classification 4
  method 72–3, 75
  properties 75
isocitrate dehydrogenase (NADP$^+$-dependent)
  enzyme classification 4
  method 72–3, 76
  properties 76
isomerases 6, 50

K_i (inhibitor constant), *see* inhibitor
K_M, *see* Michaelis constant
key enzymes 1
kinetic measurements 16–19
  see also kinetic parameters
kinetic parameters 1, 6, 7, 8, 9, 13, 14, 15, 19, 27, 29, 30, 52
  activator 9, 12
  biochemical assays 3, 11, 13, 14, 15, 26, 27, 43
  incubation time 10, 12, 15–19
  inhibitor 8, 9, 12, 13, 15, 16, 18, 22, 31, 52, 67, 84, 88
  Lineweaver–Burk plot 7–9
  maximum velocity 6–7
  metabolic fluxes 9, 29, 30
  Michaelis constant 6, 8, 9, 14, 30
  pH 9, 52
  reaction velocity 6, 7, 8, 10, 16–19
  regulation mechanisms 30, 31
  section thickness 8, 12, 17, 28, 29, 80
  specific reaction 7, 8, 12, 13, 16
  substrate concentration 6, 7, 9, 12, 14, 18, 29, 30
  substrate flux 9, 15
  temperature 9, 18
  Wilkinson plot 7, 8, 9, 14
  see also quantitative histochemistry

L-DOPA oxidase, *see* monophenol monoxygenase
lactase
  enzyme classification 5
  method 65, 67
  properties 67
lactate dehydrogenase
  enzyme classification 4
  method 72–3, 74
  'nothing' dehydrogenase reaction 103
  method 104
  properties 74
Lambert–Beer equation 28
  see also quantitative histochemistry
lead salt capture method 27, 52, 53, 55, 56, 57, 58, 59
lectin histochemistry 1
ligases 6
Lineweaver–Burk plot 7–9
  see also kinetic parameters
lipases 67
lyases 5
lysosomal membrane fragility test 102, 104, 105
  method 105

malate dehydrogenase (NAD$^+$-dependent)
  enzyme classification 4
  method 72–3, 75
  properties 75
malate dehydrogenase (NADP$^+$-dependent)
  enzyme classification 4
  method 72–3, 75
  properties 75
malic enzyme, *see* malate dehydrogenase (NADP$^+$-dependent)
malignancy 102, 103
  see also oxygen sensitivity test
α-mannosidase
  enzyme classification 5
  method 65, 66
  properties 66
marine pollution 105
  see also cellular damage
maximum velocity 6–9, 14, 30
  see also kinetic parameters
mean integrated absorbance 12, 17, 18, 28, 29
  see also cytophotometry
metabolic fluxes 9, 29, 30
  see also kinetic parameters
metal salt capture methods 17, 19, 52
4-methoxy-2-naphthylamine (MNA) 23
MIA, *see* mean integrated absorbance
Michaelis constant 6, 8, 9, 14, 30
  see also kinetic parameters
microchemistry 26, 27, 32
molar extinction coefficients 27, 28, 29
  see also quantitative histochemistry
monoamine oxidase
  enzyme classification 4
  method 87
  properties 86–7

monophenol monoxygenase
  enzyme classification 5
  method 94
  properties 93–4
multi-step reaction 22, 72, 79
  see also auxiliary enzyme
muscle diseases 102
myeloperoxidase 88
myocardial infarction 102
  see also necrosis

N-acetyl-ß-glucosaminidase
  enzyme classification 5
  method 65, 66
  properties 66
Na$^+$,K$^+$-ATPase, see ATPase
NAD$^+$-kinase
  enzyme classification 5
  method 80, 82
  properties 82
NADPH-cytochrome c (P450) reductase, see NADPH-ferrohaemoprotein reductase
NAD(P)H dehydrogenase
  enzyme classification 4
  method 85–6
  properties 22, 85
NADPH-ferrohaemoprotein reductase
  cytochrome P$_{450}$ 94
  enzyme classification 4
  method 85
  properties 84–5
NAD(P)H oxidase 61
ß-naphtholamine (NA) 23
natural chromophores 24–5, 93–5
necrosis
  'nothing' dehydrogenase reaction 103
    method 104
neotetrazolium chloride 102
  see also oxygen sensitivity test
5'-nitrosalicylaldehyde 23–4, 90–1
  see also fluorescence method
non-specific esterases 21, 67–8
  enzyme classification 5
  method 65, 68
  properties 68
'nothing' dehydrogenase reaction 102, 103
  method 104
5'-nucleotidase 19–20
  enzyme classification 5
  method 53
  properties 52–3

oligosaccharidases 65
  see also glycosidases
ornithine carbamoyl transferase 52
  enzyme classification 5
  method 57

  properties 56
ornithine decarboxylase 52
  enzyme classification 5
  method 57–8
  properties 57
oxidases 20, 21, 22, 25, 59, 61, 72, 86
oxidoreductases 3, 50
oxygen sensitivity test 102–3
  method 103

p-nitrophenyl phosphate method 54
peptidases 68–71
  methods 68–9
periodic acid-Schiff technique 24
  see also glycogen phosphorylase
peroxidase 23, 88
  enzyme classification 5
  method 89
  properties 89
peroxisomal oxidases 61
phosphatases 19, 20, 21, 25, 27, 52, 59, 64, 67
phosphofructokinase 17, 22
  enzyme classification 5
  method 80, 81–2
  properties 81–2
phosphoglucomutase
  enzyme classification 6
  method 80, 84
  properties 84
phosphogluconate dehydrogenase
  enzyme classification 4
  method 76, 72–3
  properties 76
poly(diaminobenzidine)
  molar extinction coefficient 27
polysaccharidases 65
  see also glycosidases
polyvinyl alcohol 11, 14, 25, 43, 44, 48–50, 52, 67, 105, 107
  see also tissue protectants and incubation conditions
  method 49
precision 1, 8, 9, 10, 11, 13, 20, 22, 23, 24, 25, 43, 49
  diffusion 10, 11, 23, 48, 49, 50
proteases 15, 20, 21, 23, 30, 64, 68
  method 91
proteinases 17, 68
pseudo-cholinesterases 24
pseudoperoxidases 88
purine nucleoside phosphorylase 52, 56
  enzyme classification 5
  method 80–1
  properties 80–1
pyruvate kinase 82
  enzyme classification 5
  method 80, 82
  properties 82

quantitative histochemistry 1, 7, 8, 11, 12, 13, 14, 15–19, 26, 59
  biochemical assays 3, 11, 13, 14, 15, 26, 27, 43
  cytophotometry 1, 8, 11, 14, 15, 18, 27, 28, 29
  flow cyto(fluoro)metry 1, 15, 107
  fluorescence 12, 15
  image analysis 1, 14–19, 26–30, 104
  Lambert–Beer equation 28
  mean integrated absorbance 12, 17, 18, 28, 29
  molar extinction coefficients 27, 28, 29
  units of enzyme activity 14, 26–9
  see also kinetic parameters

reaction velocity 6, 7, 8, 10, 16–19
  see also kinetic parameters
reductases 17, 21, 22, 72, 84
reliability 20, 59
  see also validity
reproducibility 1, 8, 9, 10, 11, 12, 13
resins 10, 11
  see also glycol methacrylate

section thickness 8, 12, 17, 28, 29, 80
  see also kinetic parameters
semipermeable membrane technique 11, 25, 26, 48, 50–2, 61, 92, 93
  method 51–2
  glycogen phosphorylase, method 93
  see also incubation conditions
snap freezing 36–7, 45
  method 37
specific reaction 7, 8, 12, 13, 16–19
  see also kinetic parameters
specificity 1, 2, 9–10, 11, 15
storage 45–6
  cell preparations 46
  cryostat sections 46
  frozen tissue blocks 45
  glycol methacrylate embedded tissue 46
  glycol methacrylate sections 46
substantivity 23
substrate concentrations 6, 7, 9, 12, 14, 18, 29, 30
  see also kinetic parameters
substrate flux 9, 15
  see also kinetic parameters
substrate protection 26
succinate dehydrogenase 48
  enzyme classification 4
  method 72–3, 79
  properties 79
succinate–semialdehyde dehydrogenase
  enzyme classification 4
  method 72–3, 78
  properties 78
sulfatases 19, 20, 52
synthesis reactions 24, 92
synthetases 6

test minus control reaction, see specific reaction
test reaction 7, 8, 12, 16, 18
tetrazolium salt method 17, 21–3, 27, 72–3, 79, 80, 85, 86, 87
  see also multistep reaction
thiocholine methods 24, 91, 92
tissue preparations 36–9
tissue protectants 11, 48–50, 105
  polyethylene glycol 49
  polypep 49
  polyvinyl alcohol 11, 14, 25, 43, 44, 48–50, 52, 67, 105, 107
  polyvinyl pyrrolidone 49
transferases 5, 50

UDP-glucose dehydrogenase
  enzyme classification 4
  method 72–3, 74
  properties 74
ultrastructure 36, 38
  see also electron microscopical enzyme histochemistry
units of enzyme activity 26–9
  see also quantitative histochemistry
urate oxidase 61
  enzyme classification 4
  method 63–4
  properties 63

V_{max}, see maximum velocity
validity 1, 8–13, 15, 52

Wilkinson plot 7, 8, 9, 14
  see also kinetic parameters

xanthine oxidase 61
xanthine oxidoreductase 46, 61
  enzyme classification 4
  method 86
  properties 86

zero-order velocity 6, 15
  see also maximum velocity